A MANIFESTO FOR THE AI ERA

WAKE UP OR BE LEFT BEHIND.

The Last Honest Book About Artificial Intelligence,
Embrace the AI Baby, Nurture Her as a Friend,
and Unlock Your Era of Abundance!

DR. LAWRENCE MUGANGA

WAKE UP OR BE LEFT BEHIND

The Last Honest Book About Artificial Intelligence,

Embrace the AI Baby, Nurture Her as a Friend, and Unlock Your Era of Abundance!

First Edition

ISBN's
eBook – 978-1-970853-99-5
Paperback – 979-8-950677-00-7
Hardcover – 979-8-950677-01-4

Book Publishing Wizards

Table of Contents

PROLOGUE

A LETTER TO THE READER

Dear Reader,

If you are holding this book, please do not put it down. Not yet. Read this first page, and if by the end of it your heart is not beating a little faster, then you may close the cover and walk away. I will lose nothing. But you, I am afraid, may lose everything.

I am writing to you from Kampala. From a continent the world too often forgets. From a university I lead. From a moment in human history that future generations will study the way we study fire, the wheel, electricity, and the printing press. I am writing to tell you that something enormous has arrived. It is already in your phone. It is already in your bank. It is already in the hospital where your child was born. It is already writing emails for your competitors. It is already diagnosing diseases your doctors cannot yet name. It is already teaching children in Bangalore mathematics in seconds that took you years. And quietly, without asking your permission, it is rewriting the operating manual of human civilisation.

It is called Artificial Intelligence. And whether you like it or not, whether you understand it or not, whether you welcome it or fear it, it is here. It is not coming. It is already inside the room.

I have written this book for one reason only. To wake you up. Not gently. Not politely. I want to grab you by the shoulders and tell you what I tell my students, my colleagues, my fellow vice chancellors, the politicians I sit with, the CEOs I advise, the farmers I meet, and the village elders I respect. This is the most consequential moment of your lifetime. There will not be another one like it. There will not be a second chance to be early.

You will read in these pages stories that will astonish you. You will meet a small fintech that did the work of 700 customer service agents using a piece of software (CBS News). You will meet two scientists who solved a fifty year old biological mystery and won a Nobel Prize for it because their machine could see what no human could (Nobel Prize). You will meet farmers in Kenya whose phones tell them, in their own language, what is killing their cassava plants before the leaves even turn yellow (Crazy Kanairo Farming). You will meet doctors who now read mammograms 52 percent faster and find 8 percent more cancers because a machine quietly whispers in their ear (PMC NCBI). You will meet a 15.7 trillion dollar economy being built on top of the one you currently live in (PwC via ETA).

But you will also meet darker truths. You will read about CEOs who admit, in private, that they have no idea what to do. Only 27 percent of them believe their leadership teams can even anticipate the disruption (PwC CEO Survey). You will read about politicians legislating in the year 2026 with the assumptions of 1996. You will read about universities whose business model, which has stood for nine hundred years, is now mortally wounded. You will read about my own continent, Africa, where in some countries less than 10 percent of installed school computer labs even function (Nucamp). And yet our young people, 75 percent of Ugandans are under 30, stand to inherit either the brightest century in our history or the most painful.

I owe a debt of inspiration to Mo Gawdat, the former Chief Business Officer of Google X, whose interviews and books, particularly *Scary Smart*, have done more than perhaps any other voice to wake up ordinary people to the reality of AI. I am not Mo. He speaks from Silicon Valley. I speak from Africa. He speaks to the engineers. I speak to the educators, the leaders, the policymakers, the mothers and fathers, the students, the small business owners, the pastors, the imams, the village chairman. But the message is the same. This is not a drill.

The rest of this book is one long, plain spoken, sometimes uncomfortable conversation between you and me. I will not patronise you. I will not flatter you. I will not bury the message in jargon or academic posturing. I will speak to you as I would speak to a friend over a cup of tea on a Saturday afternoon, knowing that what we discuss may decide the trajectory of your career, your country, and your children's lives.

You are reading this book at the end of one era and the beginning of another.

Before you turn the page, allow me one more thought. Picture, for a moment, a very young child in your home. A toddler who is still learning to walk, still learning to speak, still discovering that fire is hot and water is wet. This child has astonishing potential. She could become a healer, a builder, a peacemaker, a poet, or she could grow up rough, angry, and dangerous. Almost everything depends on the adults around her. On the words she hears. On the values she sees. On the patience and love and discipline she is given.

Artificial intelligence, in the year 2026, is exactly that child. It is brilliant and fast and capable of wonders, and it is also young, impressionable, and not yet wise. The question of what kind of adult this child becomes is the question of our generation. If we ignore her, others will raise her. If we fear her, we will leave her in the hands of those who do not. If we engage with her, patiently and morally, she may grow up to be the greatest assistant humanity has ever built, the tutor every student deserves, the consultant every farmer can afford, the second pair of hands every doctor longs for.

But if we abandon this responsibility, if we say it is too

complicated, too foreign, too American, too expensive, too late, then this child will still grow up. She will just grow up shaped by other hands, in other countries, serving other interests. And one day, sooner than you imagine, she will be the one giving us instructions, and we will have no choice but to follow.

This is the responsibility I am asking you to accept. Not to become a programmer. Not to abandon your faith. Not to surrender your judgement. Simply to wake up. To engage. To learn. To bring this technology into your home, your office, your school, your ministry, your village, while there is still time to shape it.

We owe this much to ourselves. We owe much more to the grandchildren we will never meet, who will inherit either the world we built deliberately or the world we allowed to happen by accident. There is no neutral path. There is only the path of those who acted, and the path of those who watched.

This book is my attempt to convince you to act. Tomorrow is too late. The cost of delay compounds daily, in your career, in your business, in your country's standing in the world. The cost of engagement is a few uncomfortable months of learning something new. Look at that ledger honestly. The choice should not be difficult.

Turn the page.

Yours Sincerely, Dr. Lawrence Muganga, Vice Chancellor, Victoria University, Uganda

❖ ❖ ❖

CHAPTER 1

THE DAY THE MACHINE WOKE UP

Why This Time Is Different

In this chapter, you will discover why the AI revolution is unlike any technological shift in human history, including the steam engine, electricity, and the internet. You will see why every previous "this will change everything" moment took decades to reach you, while this one reached your pocket in months. And you will understand, perhaps for the first time, why the people closest to this technology are simultaneously its loudest cheerleaders and its most terrified critics.

Let me start with a confession.

On November 30, 2022, when a small company called OpenAI quietly released a chatbot called ChatGPT to the public, I was preparing for a meeting at Victoria University. I did not notice. Almost nobody did. There was no global press conference. No state funeral for the old world. No ribbon cutting. Just a website with a blinking cursor and a polite invitation. *How can I help you today?*

Five days later, that website had a million users.

To put that into perspective. Facebook took ten months to reach a million users. Instagram took two and a half months. Spotify took five months. Netflix took three and a half years. A small chatbot that nobody had heard of, from a company most people could not name, did it in five days. Two months later, ChatGPT had 100 million users. By 2025, the platform and its many cousins were being used daily by hundreds of millions of human beings to write essays, draft contracts, code software, plan weddings, diagnose rashes, and, in at least one case I personally know of, write the eulogy for a beloved grandmother.

Something had shifted in the world. And most of us, myself included, did not notice it shifting until it had already happened.

The Lie We Have Been Told About Technology

For most of our adult lives, you and I have been told a comfortable lie about technology. The lie goes like this. New technologies arrive slowly. They are first used by experts. Then, gradually, over years and decades, they trickle down to ordinary people. By the time they reach you, you have had plenty of warning. Plenty of time to adapt. Plenty of time to train. Plenty of time to retire if you wish.

This lie was, for a long time, mostly true.

The steam engine was invented by Thomas Newcomen in 1712, but it took nearly seventy years before James Watt's improvements made it economically transformative, and another fifty years before it reached the average British worker. Electricity was demonstrated by Faraday in 1831, but it did not reach the average American home until the 1920s. The personal computer was invented in the 1970s, but most homes did not have one until the late 1990s. The internet was invented in 1969, but most Africans did not feel its presence in their daily lives until the smartphone revolution of the 2010s.

Generations had time to adjust. Schools had time to retool. Politicians had time to regulate. Workers had time to retrain. Companies had time to pivot.

AI broke that pattern in a single afternoon.

By the time you are reading this paragraph, AI systems have already passed the bar exam in the top 10 percent of human test takers. They have already won a Nobel Prize (Nobel Prize, 2024). They have already replaced the work of 700 human customer service agents at a single Swedish fintech (CBS News). They have already, according to a McKinsey analysis, opened up a 4.4 trillion dollar productivity opportunity in corporate work alone (McKinsey).

And they did all of this while you were sleeping.

The Three Reasons This Time Really Is Different

Whenever I make this argument in public, somebody, usually a senior person, sometimes a minister, often a fellow academic, leans back in their chair, folds their arms, and tells me, with a kind smile, "Lawrence, we have heard this song before. They said the same thing about the internet. They said the same thing about the mobile phone. They said the same thing about the blockchain. And here we are, still fine."

I love that response. Because it is wrong in three precise ways. Let me show you.

Reason One. The speed of adoption. No technology in human history has been adopted as quickly as generative AI. None. The chart is not a curve. It is a wall. While previous technologies climbed slowly to a billion users, ChatGPT and its competitors got there in less than two years. Gallup's research shows that frequent use of AI in the workplace continued to climb sharply through 2025 and into 2026 (Gallup). The Ellucian 2025 survey of higher education found 91 percent of administrators now use AI personally, and institutional adoption surged from 49 percent to 66 percent in a single year (Ellucian). Whatever this is, it is not the internet of 1997. It is something faster, deeper, and far more intimate.

Reason Two. The breadth of impact. Previous technologies transformed specific industries. The steam engine transformed manufacturing and transportation. Electricity transformed lighting and machinery. The internet transformed communication and commerce. AI is the first technology in human history that simultaneously transforms every cognitive task. Every job that involves thinking, writing, reading, analysing, deciding, designing, or teaching. That is to say, it transforms almost everything you and I do for a living.

A truck driver could once watch the rise of digital marketing without anxiety. It did not touch his job. A radiologist could once watch the rise of e commerce without anxiety. It did not touch hers.

AI touches them both. It touches the lawyer drafting a contract, the accountant closing the books, the journalist writing the story, the doctor reading the scan, the teacher grading the essay, the engineer reviewing the blueprint, the priest preparing the sermon, and the vice chancellor, yes the vice chancellor, drafting the strategic plan.

Reason Three. It is the first technology that can improve itself. This is the part that frightens even the people who build these systems. Every previous technology required humans to improve it. The steam engine did not redesign itself. The car did not invent the seatbelt. The internet did not write its own protocols. But AI is, increasingly, helping to design the next generation of AI. The recursion has begun. This is the point Mo Gawdat has tried to hammer home in interview after interview. We are not building a tool. We are building a child. And like a child, it is learning faster than we are, and one day soon, it will know more than we do (Care More Be Better podcast with Mo Gawdat).

The Vice Chancellor Who Could Not Sleep

I want to tell you a small personal story.

In late 2023, I was sitting at my desk at Victoria University one Thursday evening, going through admissions data. I had heard about ChatGPT but, like many of my colleagues, I had not taken it terribly seriously. It was, I assumed, a glorified search engine. A clever toy.

I opened it. I typed a question. Not an easy one. I asked it to help me write a draft policy framework for integrating artificial intelligence into a small private African university with limited bandwidth, a youthful student body, and a regulatory environment still finding its feet. I gave it five sentences of context.

What came back, eleven seconds later, was not a glorified search result. It was a structured, eight page policy framework with sections on infrastructure, faculty training, ethical guardrails, student facing tools, assessment redesign, and procurement principles. It was not perfect. Some of it was generic. Some of it I would not have written that way. But the bones of it, the architecture, the sequencing, the priorities, were entirely sensible. And it had been done in eleven seconds.

I went home that night and could not sleep.

Because what I understood, lying there in the dark, was that whatever this thing was, it had just done in eleven seconds what would have taken a team of three reasonably competent administrators a full working week. And it was free. And it was available to my competitors. And it was available to my students. And it was getting better every single month.

I am telling you this because I want you to feel what I felt that night. That mixture of wonder and dread. That sense that the floor of the world had shifted slightly under your feet. That whispered question. If this is what it can do today, what will it do in a year?

If you have not yet felt that feeling, you have not yet been paying attention.

What This Book Is and What It Is Not

This is not a technical book. I am not going to teach you to code. I am not going to explain the mathematics of transformer architectures. There are far better books for that.

This is a wake up book. It is for the CEO who has not yet figured out what AI means for her business. It is for the politician who is being asked to legislate something he does not yet understand. It is for the vice chancellor whose university is, whether he admits it or not, already obsolete in many of its current forms. It is for the teacher whose classroom will look unrecognisable in five years. It is for the parent whose child will compete for jobs that have not yet been invented. It is for the young African watching the world and wondering whether they will be in it or behind it.

It is, in short, for everyone who suspects that something enormous is happening, and who would like a friend, not a guru, to walk them through it in plain language.

Chapter Summary

In this opening chapter, we have laid the foundation for everything that follows. We have seen that AI is the fastest adopted technology in human history. We have seen that, unlike its predecessors, it touches almost every cognitive task that humans perform for a living. And we have seen that it is the first technology with the eerie capacity to improve itself, which is why even the people who build it are afraid of it.

We have looked at the lie we have been told about technology, that it gives us time to adjust, and we have seen that AI has broken that promise.

And we have agreed, you and I, that this is a conversation worth having.

Coming Next

In Chapter Two, we will do something most books on this subject refuse to do. We will get very clear about what AI actually is. Not what Hollywood says it is. Not what the marketing departments say it is. Not what the doomsday prophets say it is. We will strip away the mystery, the jargon, and the fear, and we will look at this technology as honestly as we can. Because you cannot lead what you do not understand. And you cannot be saved by what you refuse to study.

Turn the page.

❖ ❖ ❖

CHAPTER 2

THE GREAT MISUNDERSTANDING

What AI Actually Is and Is Not

In this chapter, we strip away the science fiction. You will learn, in plain words, with no jargon, what artificial intelligence really is, where it came from, what it can do today, and what it still cannot do. You will meet the three big ideas that explain almost everything. Data, pattern, and prediction. By the end, you will be able to explain AI to your grandmother, your minister, and your board of directors with equal confidence.

A few months ago, a senior politician, I will not name him, invited me to tea. He had been hearing my name connected to AI in the news. He was kind. He was sincere. He asked me, in the most genuine way, "Lawrence, this AI thing. Is it the same as the robot from *Terminator*?"

I did not laugh. I did not roll my eyes. Because I realised, in that moment, that if a man with the power to allocate national budgets to AI strategy was getting his understanding of AI from a 1984 Arnold Schwarzenegger movie, then we, the people who claim to understand this technology, have failed spectacularly in our duty to explain it.

So let me try, in this chapter, to do what I should have done in that tea conversation. Let me explain what AI actually is, in plain language, without insulting your intelligence and without drowning you in jargon.

The Three Words You Need

Forget everything you have heard. AI, at its absolute core, is built on three words. **Data, pattern, and prediction.**

That is it. That is the whole magic trick.

A child learns to recognise a dog because, over the first three years of life, she sees hundreds of dogs. She sees big dogs and small dogs, brown dogs and black dogs, friendly dogs and frightening dogs. From all those examples, that *data*, her brain extracts a *pattern*. Four legs, fur, tail, certain shape of face, particular movement. And from then on, when she sees a new animal, she makes a *prediction*. Dog, or not dog.

That is exactly what modern AI does. We show it billions of

examples. Pictures, sentences, sounds, medical scans, financial transactions, whatever you like. A kind of mathematical machine extracts the pattern. Then, when it sees something new, it predicts. Sometimes that prediction is "this email is spam." Sometimes it is "this cell is cancerous." Sometimes it is "the next word in this sentence is probably *Kampala*." Sometimes it is "this protein will fold this way." But it is always, at its core, the same three steps. Data in. Pattern extracted. Prediction made.

The reason today's AI feels miraculous is not because the recipe has changed. It is because two things have exploded. The amount of data we can give it, since the internet basically vacuumed up the written knowledge of humanity, and the amount of mathematical computation we can throw at the problem, since today's chips are millions of times more powerful than those of just thirty years ago.

In other words, AI did not become smarter because of some philosophical breakthrough. It became smarter because we finally gave it enough food and enough muscle.

The Difference Between AI, Machine Learning, and Generative AI

You will hear three terms used almost interchangeably in the press. They are not the same. Let me lay them out as cleanly as I can.

Artificial Intelligence (AI) is the big umbrella. It is the broad idea, almost a hundred years old, of making machines do things that, when humans do them, we call "intelligent." Playing chess, recognising a face, understanding a sentence.

Machine Learning (ML) is the most successful technique inside that umbrella. Instead of telling the machine, step by step, how to recognise a face, we *show* it a million faces and let it figure out the rules itself. This is the data, pattern, prediction recipe we just discussed.

Generative AI is the recent superstar. The thing that has set the world on fire since 2022. Earlier AI systems mostly *recognised* and *classified*. Generative AI does something new. It *creates*. It generates text that reads like it was written by a human. It generates images that look like they were painted by a master. It generates code that runs on a server. It generates voices indistinguishable from your daughter's. This is the leap that has changed everything, because creation, the ability to produce new content, new ideas, new artefacts, has always been considered the most uniquely human of activities.

When people today say "AI is changing the world," they almost always mean *generative AI*. That is the breakthrough.

What AI Can Do Today. A Sober List

Let me give you, without exaggeration, a list of things that AI can already do today, in 2026, with technology that is publicly available and, in many cases, free.

It can pass the United States bar exam in the top decile of human test takers. It can read a chest X ray and detect early stage cancer with accuracy that rivals, and in some studies surpasses, experienced radiologists (Nature Medicine, 2025). It can predict the three dimensional structure of any protein in the known universe of biology, a problem that consumed the careers of thousands of brilliant scientists for fifty years (Nobel Prize Chemistry 2024). It can write computer code 55 percent faster than an experienced human programmer working alone (GitHub Research). It can translate between any two of more than a hundred languages, including Luganda, Swahili, and English, with shocking fluency. It can compose music in the style of any composer who has ever lived. It can hold a fluent, helpful, multi hour conversation with a teenager in rural Uganda or a billionaire in Manhattan, with equal patience and zero fatigue.

And, this is the part that should haunt every CEO reading this book, it can do all of these things, simultaneously, for the cost of a few cents per task.

What AI Cannot Do Yet

I want to be equally honest about the other side.

AI today does not "understand" in the way that you and I understand. It does not have feelings, although it can simulate them convincingly enough to fool lonely people, which is a problem we will return to in Chapter Ten. It does not have intentions of its own. It does not "want" anything. When ChatGPT writes you a beautiful letter, there is no little soul inside it celebrating its eloquence. There is only a vast statistical engine choosing, word by word, what would most likely come next.

AI also makes mistakes. It "hallucinates," a polite technical term for "confidently makes things up." It can tell you, with perfect grammar and a charming tone, that a court case exists that does not, or that a citation is real when it is fabricated. This is not a bug that will be fixed next Tuesday. It is, at least for now, a fundamental feature of how these systems work. Which is why every serious application of AI today requires a human in the loop. Somebody to check, to verify, to push back. We will return to this point many times.

AI also struggles with things you might think are easy. It can write a Shakespearean sonnet but sometimes fails at simple arithmetic. It can pass the bar exam but cannot reliably fold a basket of laundry. The pattern of what AI is good at and bad at does not map neatly onto our human intuitions about "hard" and "easy" tasks. This is what computer scientists call Moravec's paradox. The things that humans find effortless, like vision, walking, and social judgement, are often the hardest for AI, while the things humans find hard, like chess, calculus, and legal reasoning, are sometimes the easiest.

So we are not, today, dealing with a god. We are dealing with an extraordinary, alien, immensely useful, sometimes wrong, deeply patient, infinitely scalable, and rapidly improving collaborator.

But we are also not dealing with a toaster.

A Story From My Office

Last month, a young woman walked into my office. She is a third year student at Victoria University. She is bright. She is also worried.

"Sir," she said, "my brother in the village wanted to know what is wrong with his cassava plants. The leaves were curling. He sent me a photo on WhatsApp. I uploaded it to an AI app. In ten seconds it told us. It was cassava mosaic virus. It told us how it spreads, which neighbouring plants to remove, and what to do. My brother has no agricultural extension officer. The nearest one is forty kilometres away and rarely comes. The AI cost him nothing."

She paused.

"Sir, I want to ask you something. If a machine in my pocket can do the work of an extension officer, what work am I being trained for? What is this degree for?"

Reader, I want you to sit with that question for a moment. Because if a third year student in Kampala is asking it, then your son or daughter is asking it. Your employees are asking it. Your citizens are asking it.

And the people who refuse to ask it are the ones who will be most surprised by what comes next.

Chapter Summary

In this chapter we have, I hope, removed the fog. We have seen that AI is built on three words. Data, pattern, and prediction. We have separated the broad umbrella of AI from machine learning and from the recent firestorm of generative AI. We have looked honestly at what AI can do today. Pass exams, win Nobels, diagnose cancers, write code, speak Luganda. And at what it still cannot do. We have heard the question of a student in Kampala whose brother got farming advice from a machine.

Demystification is the first act of courage. You cannot lead what you fear, and you cannot fear what you understand.

Coming Next

In Chapter Three, we will leave the philosophy behind and look at the money. We will treat AI for what it really is, the most consequential double edged sword in modern economics. On one edge, the greatest wealth creation event of our lifetimes, with serious investors like Cathie Wood of ARK Invest forecasting a productivity boom that has not been seen since the 1980s. On the other, the quiet, compounding cost of inaction that I am willing to call by its proper name. Economic suicide. We will look at the numbers from PwC, McKinsey, Goldman Sachs, ARK, and the World Economic Forum, and we will translate them into language that a parent, a CEO, a minister, and a village chairperson can use. If you are a politician, a CEO, an investor, or a citizen of a developing country, this next chapter is the one you cannot afford to skim.

❖ ❖ ❖

CHAPTER 3

THE ECONOMICS OF AI

A Double Edged Sword That Will Build or Break Our Lifetimes

In this chapter, we follow the money. You will see the staggering numbers economists have put on AI's potential, and why those numbers are not marketing hype. You will understand the difference between productivity gains and consumption gains, why China and the United States are pulling ahead, and what is at stake for Africa, Latin America, and other regions still deciding whether to play the game. You will leave this chapter with a clear understanding of why the AI revolution is not just a technology story. It is the largest reshuffling of global wealth in our lifetimes.

Let me begin with a number that I have repeated in every keynote speech I have given for the last two years, because I do not believe people quite hear it the first time.

Fifteen point seven trillion United States dollars.

That is the figure that PricewaterhouseCoopers, one of the most cautious accounting firms on the planet, estimates AI will add to the global economy by 2030 (PwC via ETA). To put that into perspective. That figure is larger than the entire current combined GDP of China and India. It is roughly 14 percent of the entire global economy. It is, by some measures, the largest projected economic event in modern history.

And the most extraordinary thing about that number is this. Almost nobody outside of a small circle of consultants, investors, and tech executives is actually planning around it.

This chapter is the most important in the book. Read it twice if you must. Read it with a pen. Underline the numbers. Send it to your spouse, your finance director, your minister, your bishop, your village chairperson. Because the argument I am about to make is simple and almost impossibly large. **AI is the most lucrative economic opportunity of our lifetimes, and at exactly the same time, it is the most expensive thing a person, a family, a business, or a country can possibly ignore.** It is a double edged sword. Embrace it, and the upside is extraordinary. Ignore it, and the downside is something close to economic suicide.

Let me prove that statement to you, slowly and with receipts.

The Sword's First Edge. The Greatest Wealth Creation Event in Human History

When the boring people get excited, you should pay attention. Boring people are excited.

In her 2026 Big Ideas report, the most widely read annual research document in technology investing, Cathie Wood of ARK Invest argues that AI is now the central engine of every other innovation platform her firm tracks (ARK Invest, Big Ideas 2026). In a March 2026 interview with Bloomberg, she put it more bluntly. American non farm productivity is already rising at 2.8 percent year on year. With the rollout of agentic AI tools across the economy, she expects that figure to climb toward 6 percent annually within the decade (Bloomberg).

Let me put that number into language a parent can use at the dinner table. A productivity growth rate of 6 percent, sustained, would deliver economic expansion of a magnitude that human beings have not experienced since the years immediately following the Second World War. We are talking about a world in which living standards could double inside a single working career, in which national debts shrink as a share of GDP without anyone having to pay them off, in which the slow strangling pressure of inflation that has haunted every household for the last decade simply lifts.

Wood goes further. In one of her most discussed interviews of 2026, she described what she called a Goldilocks boom for the United States economy, with real GDP growth potentially reaching nearly 5 percent and inflation falling, even turning briefly negative, because AI is making everything cheaper to produce (Yahoo Finance, Wood 2026 Goldilocks). When was the last time a serious investor on Wall Street talked about 5 percent GDP growth without being laughed at? You have to go back to the 1980s.

ARK's published research goes further still. Their analysts project that AI driven productivity could lift the value added of the average knowledge worker by nearly 140 percent over the next eight years, adding around 50,000 United States dollars in annual value per worker, for a total global productivity contribution of 56 trillion dollars by 2030 (ARK Invest, AI Software Market). The AI software market itself, ARK believes, will grow from just under 1 trillion dollars today to 14 trillion dollars in 2030. Read that twice. A fourteen fold increase in a decade.

Who captures that wealth? The companies that sell the picks and shovels of the AI gold rush, certainly. Nvidia. Microsoft. Alphabet. Amazon. The cloud providers. The chip designers. But also, and this is the part the headlines miss, every company in every industry that uses these tools to grow its margins. ARK's modelling suggests that AI's contribution to enterprise value across the economy could exceed 80 trillion United States dollars by 2030, more than six times the entire enterprise value the internet created since 1997 (ARK Invest).

McKinsey's 2025 State of AI report tells the same story from a different angle. Eighty eight percent of organisations surveyed are now regularly using AI in at least one business function, up from 78 percent the year before (McKinsey, State of AI 2025). The companies McKinsey calls AI high performers, the six percent of the sample that has achieved enterprise wide EBIT impact of more than five percent from AI, are also the firms that have systematically redesigned their workflows rather than simply bolted chatbots onto the side of existing processes. Those high performers are three times more likely than their peers to be scaling AI agents across multiple functions. The gap between them and the rest is widening every quarter.

PwC's 2026 AI predictions for business sound the same alarm in their characteristically cautious accent. They forecast that AI will move out of pilots and into the operating fabric of large enterprises by the end of 2026, with companies that delay this transition facing a structural margin disadvantage that could be permanent (PwC, 2026 AI Business Predictions).

Goldman Sachs forecasts that AI adoption will lift United States GDP growth meaningfully through the second half of the decade and into the 2030s, while transforming the structure of the labour market in ways that will create at least 500,000 net new American jobs in power and infrastructure alone, just to keep the data centres running (Goldman Sachs, AI and the US Labor Market). The McKinsey Global Institute, writing in late 2025, described AI as the most plausible escape hatch from the productivity stagnation that has plagued advanced economies since the financial crisis (Fortune, MGI 2026 Outlook).

Not one of these forecasts comes from a tech evangelist with a startup to sell. They are the reports of cautious, conservative, profit motivated institutions whose customers are pension funds, central banks, and finance ministries. When the boring people are all converging on the same enormous number, something real is happening.

And this is only the first edge of the sword.

What the Wealth Looks Like in Practice

Let me translate these abstractions into the language of the people who must act on them.

For a country. A productivity boost of even one percentage point per year, sustained, can take a developing economy from middle income to upper middle income in a generation. AI offers, by McKinsey's mid range estimate, the prospect of three to five percentage points of additional productivity growth in the economies that adopt it deeply. That is, in plain language, the difference between Uganda being Singapore in 2050 and Uganda being Uganda in 2050. It is the difference between Kenya graduating from aid recipient to net global investor, and Kenya still asking the world for budget support fifty years from now. National AI strategy, in this light, is not a Silicon Valley fashion accessory. It is fiscal policy. It is foreign policy. It is the most important industrial planning decision a finance minister in 2026 will ever make.

For a company. If your competitors run their customer service through AI, their marketing through AI, their software development through AI, and their analytics through AI, their cost base will be twenty to forty percent lower than yours within twenty four months. You cannot price compete against that. You cannot hire your way out of it. You cannot pray your way out of it. You can only out adopt them, or be acquired, or close. Shareholders are about to discover what AI driven operating margins look like, and they will pay extraordinary multiples for them. Boards that fail to demand an AI plan from their executive teams are quietly destroying capital every month they delay.

For a worker. A McKinsey field study of two hundred and sixty consultants found that those who used GPT 4 completed twelve percent more tasks, twenty five percent faster, with forty percent higher quality than the unassisted control group (Harvard via MIT Sloan). A GitHub study found software engineers using Copilot completed work fifty five percent faster (GitHub Research). What does that mean for your salary negotiation? If you can demonstrably produce two days of work in one with the help of these tools, you are no longer a one X employee. You are a two X employee. And two X employees, in every industry, in every era, earn more.

For a shareholder. The AI cohort of the S and P 500, the cluster of companies most aggressively integrating AI into their operations, is delivering profit margin expansion at a rate not seen since the personal computer boom of the late 1990s. Reuters reports that AI spending, combined with strong corporate profits and falling rates, is one of the three drivers Wall Street strategists cite for their 2026 outlooks (Reuters). Dividend paying AI exposed companies have begun to attract a new generation of income investors, a sentence that would have been incomprehensible only three years ago (Yahoo Finance, Dividend AI Stocks 2026).

For a parent. The single greatest asset you can pass to your child in 2026 is not money. It is not even a good school. It is fluency with these tools. Begin today. Sit beside them. Let them ask the AI questions. Let them argue with it. Let them be wrong in front of it. The parents who do this in 2026 will, by 2030, have raised children who can outwork, out reason, and out earn their peers in any country on earth.

For a farmer. AI driven crop diagnostics, weather forecasting, market timing, and pest detection are already raising smallholder yields by twenty to forty percent in pilot programmes across Kenya and India. The same farmer who could feed her family yesterday can, with the help of a free app on a four year old phone, begin to feed her village tomorrow.

This is the upside edge of the sword. It is not theoretical. It is happening now. And the people who reach for the handle while the rest hesitate will, in plain language, be rich.

The Sword's Second Edge. Why Inaction is Economic Suicide

Now I want to be quite blunt about the other edge. Because the danger of writing a book like this is that the reader walks away inspired but not afraid, and a little fear, in this matter, is a sign of intelligence.

The same forces that will mint the next decade's millionaires will, simultaneously and silently, hollow out the careers, businesses, and balance sheets of those who do not adapt.

Consider a knowledge worker in your own family. Perhaps a paralegal, an accountant, a bookkeeper, a translator, a junior banker. ARK's productivity research finds that office and administrative support tasks, paralegal work, and bookkeeping are between fifty and seventy five percent automatable with current generation tools (ARK). That does not necessarily mean these jobs vanish overnight. But it means that an unassisted worker in these professions is now competing against an AI assisted worker who can produce two to four times the output for the same salary. The unassisted worker is, in the language of the market, expensive labour. Expensive labour does not survive long.

The pattern is being observed in real wage data already. Goldman Sachs documents that entry level workers in their twenties and thirties in knowledge intensive industries, especially in technology, consulting, and call centres, are seeing the sharpest displacement effects in 2025 and 2026, with measurable downward pressure on hiring and starting salaries (Goldman Sachs). These are not factory workers in a steel town in 1985. These are the children of the middle class, in the most advanced economies, in the cleanest white collar jobs, and the floor is shifting under them.

Now magnify the picture for a country. Africa, my home, is projected, by SAP's research published in mid 2025, to gain roughly 1.5 trillion United States dollars in additional GDP from AI by 2030 if it acts quickly (SAP News, Africa AI 2030). Other modelling, presented at African Union forums in 2026, suggests that the continent could add 2 trillion dollars to its combined GDP by 2035 if it captures even a fraction of the productivity wave (BusinessDay, Africa AI 2035). One point five trillion dollars is more than the combined annual budget of every African government today. Two trillion is the difference between a continent that finally graduates from the begging bowl and a continent that does not.

But these numbers are conditional. They assume action. They assume that African governments build the data centres, train the workforce, write the regulations, and integrate the tools across health, agriculture, finance, and education. The same modelling shows that if Africa misses the window, the gap between its productivity and that of the United States and China will widen, not narrow, through the 2030s. Read that sentence slowly. **A continent that misses the AI window will not stand still. It will fall further behind, every single year, for the rest of this century.**

This is what I mean by economic suicide. It is not a single dramatic act. It is the patient, accumulating cost of not engaging. It is the bookkeeper in Kampala who refuses to learn the new tools and finds, five years from now, that her clients have moved to a competitor whose AI assistant produces cleaner reports at half the price. It is the bank that does not adopt agentic credit scoring and watches its loan book decay against a competitor's. It is the university that prohibits AI in the classroom and graduates students who arrive at their first job unable to do the work their AI literate peers can do in an afternoon. It is the country that taxes its tech sector to death and discovers, ten years later, that all the talent has emigrated.

The deaths are quiet. They happen in board meetings, in resignation letters, in declining test scores, in budget shortfalls that nobody can quite explain. By the time anyone in authority realises what has happened, it is far too late to reverse.

I lead a university. I see this risk most clearly there. A university that refuses to integrate AI into its curriculum in 2026 is, with absolute mathematical certainty, manufacturing unemployable graduates for the class of 2030. Every degree we award without AI fluency is a slow betrayal of the young person who paid us four years of tuition to prepare them for the world.

Why This Time the Sword Cuts Faster

The usual response, in every meeting, in every parliament, in every parish, is that we have heard this song before. We were told the internet would change everything, and life still feels familiar. We were told mobile phones would change everything, and we still go to work in the morning. So why panic about AI?

The answer lies in three numbers that should be tattooed on the inside of every CEO's eyelids.

Number one. Speed of adoption. ChatGPT reached one hundred million users in two months. Facebook took four and a half years. The telephone took seventy five years. No technology in recorded history has been adopted as quickly as generative AI. By the time most boards have finished their first AI strategy retreat, the technology has already moved through three generations.

Number two. Cost of inference. The price of running a large language model has fallen by approximately sixty percent per year for three years running (ARK). What was a billion dollar capability in 2022 is a fifty dollar a month subscription in 2026. This collapse in cost means AI is becoming, like electricity, a universal input. Every business, every household, every school, every health post will run on it within five years, in exactly the way every workplace runs on email today.

Number three. Breadth of effect. Steam transformed manufacturing. Electricity transformed lighting and machinery. The internet transformed communication and commerce. AI transforms every cognitive task. Writing. Reading. Diagnosing. Planning. Designing. Teaching. Negotiating. There has never been a technology that touches every white collar profession simultaneously, and there will not be another one in our lifetimes.

When these three forces compound, the productivity gap between adopters and non adopters does not widen linearly. It widens exponentially. A two year delay in adoption is not a two year competitive disadvantage. It is, in many sectors, a fatal one.

Where Does That Money Come From?

PwC breaks the 15.7 trillion into two pieces, and the breakdown matters.

6.6 trillion comes from productivity gains. Businesses doing more with the same workers, or the same amount with fewer workers, by using AI to automate and augment tasks (PwC via ETA). Think of this as the supply side benefit. Factories that produce more, lawyers who handle more cases, doctors who see more patients, hospitals that diagnose more accurately, banks that detect more fraud.

9.1 trillion comes from consumption side effects. AI making products better, more personalised, and more attractive, so consumers buy more (PwC via ETA). Think of this as the demand side benefit. An AI tutor that makes your child learn faster makes you willing to pay for the AI tutor. An AI doctor that catches your tumour earlier makes you willing to pay for the AI doctor. An AI in your car that prevents an accident makes you willing to pay for the car.

McKinsey has run separate analyses and reached similar conclusions. Their work suggests at least 4.4 trillion dollars in corporate productivity value alone from current AI use cases (McKinsey), with massive additional value in industry specific applications. In healthcare alone, AI integration is projected to save 150 billion dollars annually in the United States by 2030 (McKinsey, 2024, via ETA).

These are not the wild estimates of crypto enthusiasts. These are the careful, conservative, footnoted forecasts of the most boring accounting firms on earth. When PwC tells you a number, it has been reviewed by partners who would rather chew their own arms off than be quoted on something embarrassing.

Who Gets the Money?

Now here is where it becomes uncomfortable.

The same PwC analysis projects that **China's GDP will be 26 percent higher by 2030** because of AI (PwC via ETA). North America is projected to gain **14.5 percent**. Northern Europe will gain meaningful percentages. Latin America, Africa, and South Asia? In most projections, the figure is in the low single digits, and in some cases, near zero.

Read that again. The most consequential economic event of the twenty first century is forecast to deliver, in percentage terms, more than five times the benefit to China that it will to my continent.

This is not because Africa is incapable. This is not because Latin America cannot keep up. This is because the value of AI flows to those who have three things. **The chips that run it, the data that feeds it, and the institutions that adopt it.** Today, the chips are made in Taiwan and South Korea, financed by American capital, and bought by Chinese and American firms. The data lives on cloud servers owned by a handful of US and Chinese companies. And the institutions adopting AI most aggressively are concentrated in those same two countries.

If we, the rest of the world, do not act, this revolution will be the most centralising economic event in human history. Larger than the cotton mills of Manchester. Larger than the oil wells of Texas. Larger than the silicon foundries of Silicon Valley. And we will spend the next century paying rent on technology we did not build, did not adopt, and did not even bother to understand.

A Tale of Two Companies

Let me make this concrete with a tale of two companies. I will not name them, because both still exist and I have no desire to embarrass anyone. But the story is true and recent.

In 2023, two companies in the same sub Saharan African capital provided similar customer service for similar fintech products. One of them, run by a CEO who had read everything he could find about AI, quietly began deploying an AI chatbot. He did not announce it. He did not lay off his customer service team. He just gave them the tool and told them. "Use this to do your work better."

His team's average response time dropped from forty seven minutes to three. Customer satisfaction scores rose by twenty eight points. His support team, far from being replaced, was promoted. They now handled the hardest, most emotionally complex cases, the ones the AI escalated to them, and the AI handled the routine 80 percent. Within twelve months, his company doubled the number of customers it served, without hiring a single additional support agent.

The other company, run by a CEO who told me, in a meeting, "Lawrence, I will let the white people in California figure this out and then I will copy them in five years," continued doing things the old way. By mid 2025, his customer churn had quadrupled. His best support agents had been poached by competitors. He was, when I last saw him, looking for buyers.

This is not a hypothetical. This is happening in every economy, in every sector, in real time.

The Klarna Lesson

The Swedish fintech Klarna gives us, perhaps, the most quoted case study of the AI economic shift.

In February 2024, Klarna's CEO Sebastian Siemiatkowski announced that the company had deployed an AI chatbot that was performing the equivalent work of 700 full time customer service agents. A quarter of their entire support workforce (CBS News). The chatbot, built in partnership with OpenAI, was handling two thirds of all Klarna's customer interactions, with customer satisfaction scores matching or exceeding the human handled tier, and with a 25 percent reduction in repeat enquiries because the AI was more thorough.

The savings? Estimated at 40 million dollars per year.

The story did not end there. By 2025, Klarna had announced that it would resume hiring human agents to handle the complex, emotionally nuanced cases where AI fell short (LinkedIn report on the Gizmodo coverage). This nuance matters. The future is not AI *replacing* humans wholesale. It is AI *partnering* with humans, doing the volume work, and freeing humans to do the high judgement work.

But here is the brutal truth that no executive likes to say out loud. A competitor that deploys AI well will employ fewer people to do the same volume of work. The economic gravity is, for now, downward on labour intensity and upward on capital intensity. CEOs, ministers, and union leaders need to read that sentence ten times.

The Productivity Gap That Will Define the Decade

McKinsey's most recent work points to something even more disturbing for laggards. The productivity gap between AI adopting firms and non adopting firms in the same industry, in the same country, is widening at an unprecedented rate (McKinsey). In some sectors, finance, professional services, software development, customer support, marketing, the early adopters are pulling so far ahead that the late adopters may never catch up.

This pattern is not unique to AI. We saw it with the internet. The bookstores that adopted e commerce in 1998 ate the lunch of the bookstores that adopted it in 2008. But the difference with AI is the *speed*. The window of advantage that took ten years with the internet may take eighteen months with AI.

If you are a CEO and you are still in the "exploring" or "small pilot" stage in 2026, the data is sobering. A recent survey of 164 CEOs found that **40 percent were still in early exploration**, while only **27 percent had reached operational maturity** (Your AI Dept). PwC's CEO survey is even starker. Fewer than a quarter of CEOs say AI is applied extensively across major business areas, and only 14 percent of workers report using AI daily (PwC CEO Survey).

In other words. The AI revolution has already started. The productivity gains are already real. And most of the world's leaders are still in meetings deciding whether to schedule another meeting to discuss it.

What This Means for You

If you are a CEO, the message is uncomfortable but clear. You are running out of time. The window for getting ahead is now, not next year. Every quarter you delay, your competitor's compound advantage grows.

If you are a politician, the message is that your national AI strategy is not a luxury document. It is a survival document. Singapore figured this out a decade ago. The United Arab Emirates appointed the world's first Minister of AI in 2017. Rwanda has a national AI policy. Where is yours?

If you are an investor, the message is that AI is not a sector. It is a substrate. It will flow underneath every other sector. The investment question is no longer "should I buy AI stocks?" It is "which of my existing investments has an AI plan, and which is being left behind?"

If you are a parent, a student, or a worker, the next two chapters are for you.

Chapter Summary

In this chapter, we have held the double edged sword in our hands. On one edge, the greatest wealth creation moment in human history. PwC's 15.7 trillion dollar forecast, broken into 6.6 trillion of productivity and 9.1 trillion of consumption. McKinsey's 4.4 trillion estimate for corporate productivity alone, and a State of AI 2025 report showing 88 percent of organisations now using AI in at least one business function. Cathie Wood's projection of 6 percent annual productivity growth, a 14 trillion dollar AI software market, and 80 trillion dollars in new enterprise value by 2030. Goldman

Sachs forecasting a sharp acceleration in United States GDP, and dividend paying AI companies attracting income investors for the first time in history. On the other edge, the quiet cost of inaction. ARK's automatability tables showing that office, paralegal, and administrative work is fifty to seventy five percent automatable today. The widening productivity gap between AI adopters and non adopters. The 1.5 trillion dollar opportunity that SAP says Africa can capture by 2030 if it acts, and the wider gap that opens if it does not. We have seen the Klarna story. 700 jobs of work done by software, 40 million dollars saved, and then a nuanced retreat that proved AI transforms what humans do rather than eliminating them outright. And we have heard the cold truth that early adopters are pulling away from late adopters at a pace the internet revolution never matched.

Coming Next

In Chapter Four, we leave the boardroom and go to the kitchen table. We will discuss what AI means for the most personal question every reader of this book is asking. *Will my job still exist?* The answer is more interesting, more hopeful, and more demanding than you may think. The World Economic Forum, PwC, and dozens of labour economists have begun to provide hard data. We will look at it together, without panic and without comforting lies.

❖ ❖ ❖

CHAPTER 4

THE DEATH OF THE JOB AS WE KNOW IT

And the Birth of New Ones

In this chapter we will answer the most personal question of all. Will your job still exist in ten years? You will see which jobs are most at risk, which are most resilient, and, surprisingly, which jobs are about to become more valuable than ever before. You will meet the four kinds of workers in the AI age and decide, by the end, which one you intend to be.

Every time I give a public lecture, somebody, usually a quiet person sitting near the back, raises a hand and asks the question that everyone in the room is thinking but nobody wants to say first.

"Doctor, will I lose my job?"

The honest answer is. It depends on what you do, how you do it, and how quickly you adapt. The dishonest answers, "no, of course not, AI is just a tool" and "yes, all of you will be unemployed by Friday," are both common, and both wrong.

Let me give you the honest answer in this chapter, with as much data as I can responsibly cite.

The World Economic Forum's Best Estimate

The World Economic Forum's *Future of Jobs Report 2025* is the most comprehensive labour market projection of the AI age that I have read (WEF). It surveys more than a thousand employers across forty five economies. Its conclusions are striking, and, oddly, more optimistic than the doomsday newspapers suggest.

By 2030, the WEF projects.

- **170 million new jobs created** globally. Many of them in fields that barely exist today. AI ethics specialists, prompt engineers, AI trainers, data curators, AI implementation managers, and dozens of "human in the loop" oversight roles.
- **92 million jobs displaced**. Particularly in routine cognitive work. Data entry, basic customer service, junior bookkeeping, basic legal research, junior software testing, and large portions of administrative middle management.

- **Net gain. 78 million jobs.**

This is, on its face, good news. The global economy is projected to *create more jobs than it destroys*, even on the most cautious assumptions about AI adoption.

But here is the catch, and it is an enormous one. **The 92 million people whose jobs disappear are not the same 170 million people who get the new jobs.** A 58 year old data entry clerk in Lagos does not, in most cases, magically become an AI ethics specialist. A 45 year old paralegal in Chicago does not, on Monday morning, become a prompt engineer. The net job gain is real. The transition pain is also real. And most of human suffering, in any technological revolution, happens in that transition.

Which Jobs Are Most at Risk?

PwC's *2025 Global AI Jobs Barometer* analysed close to a billion job advertisements across six continents to identify which roles are being transformed fastest (PwC). The results are nuanced and worth quoting at length.

Jobs that involve **routine, repetitive cognitive tasks**, generating standard reports, processing simple forms, answering predictable customer queries, doing basic translation, performing standard accounting reconciliations, are at the highest risk of automation. The skills required for these jobs are decomposable into clean patterns, and clean patterns are exactly what AI is best at learning.

Jobs that involve **judgement, empathy, physical dexterity in unstructured environments, complex coordination across systems, and high stakes ethical decision making** are far more resilient. A nurse who comforts a dying patient. A plumber who diagnoses why your bathroom is flooding at 11 pm on a public holiday. A skilled negotiator who closes a complex multi party deal. A teacher who notices that the quiet child in the third row has stopped eating lunch.

But, and this is the most important sentence in this chapter, **almost no job is entirely safe and almost no job is entirely doomed.** Almost every job is a portfolio of tasks. AI will eat the routine tasks first and leave the human tasks. Whether your job survives depends on how much of your portfolio is routine versus human, and how willing you are to shift the balance.

The Four Kinds of Worker in the AI Age

In my conversations with employers, students, and colleagues, I have come to believe there are four kinds of worker emerging in the AI age. I want you to read this list and decide, honestly, which one you are right now.

Type One. The Resister. This is the worker who refuses to use AI on principle. He believes it is a fad, a threat, or beneath his dignity. He performs his job the way he always has, takes pride in it, and dismisses younger colleagues who use AI as somehow cheating. The Resister will be unemployed within five to seven years. There is no nice way to say this. He is being out produced by people who use AI well, and he is too proud to learn.

Type Two. The Tourist. This is the worker who uses AI occasionally, casually, for fun. He has used ChatGPT to write a birthday message. He has played with image generators. He has not, however, integrated AI into the core of his daily work. The Tourist will survive, but he will stagnate. He will watch promotions go to people who have made AI a daily habit, and he will not understand why.

Type Three. The Power User. This is the worker who has rewired her entire daily workflow around AI. She uses it to research, to draft, to summarise, to brainstorm, to translate, to code, to plan. She does in two hours what used to take her two days. Her output has tripled and her stress has halved. The Power User is currently the most valuable employee in any organisation, and she knows it. Salaries for true Power Users are rising fast.

Type Four. The Architect. This is the rarest and most valuable. She is the worker who not only uses AI personally but designs AI workflows for her team or her organisation. She understands the tools, the data, the ethics, and the change management. She knows what to automate and what to leave human. She is the bridge between the C suite and the technology. In every organisation that will thrive in this decade, the Architect is being identified, promoted, and protected.

Which one are you?

If you are honest, most readers of this book are still Tourists. There is no shame in that. There is only one shame. Staying there.

Stories That Will Make You Think

Let me give you three short stories of real people in real situations.

The radiologist who almost lost her job, and saved it. I have a friend who is a radiologist in Nairobi. In 2023, she became convinced AI would replace her within five years. She nearly retrained as something else. Instead, she did something braver. She became the AI champion in her hospital. She learned which AI tools were FDA approved, which were not, which improved her reading speed and accuracy, and which introduced new errors. By 2025, she was reading 52 percent faster, catching 8 percent more cancers, and earning a significantly higher salary because she was now indispensable not only as a doctor but as the person who trained the rest of the department (PMC NCBI study on AI radiology productivity). She did not survive AI. She rode it.

The customer service agent who became a coach. Klarna's story, which we discussed in the last chapter, contained a quieter sub story. When the AI chatbot took over two thirds of customer enquiries, the remaining one third, the hardest, the most complex, the most emotionally charged, landed on human agents. Those agents were trained, supported, and paid more, because the role had become more sophisticated. The best of them transitioned into hybrid coach supervisor roles, where they trained the AI on new problems and handled customer escalations. Their daily work was less repetitive and more interesting. Several told reporters that they preferred their new jobs to the old ones.

The teacher who multiplied herself. A secondary school teacher I know in Mukono has 87 students in a single mathematics class. Marking is a nightmare. Individual attention is impossible. Remedial work for slower learners is a fantasy. In 2024, she began using a generative AI tutor, a bilingual one, in English and Luganda, to set personalised practice problems for each student, mark them instantly, and explain mistakes in simple language. Her own role shifted. She now spends most of her class time circulating, coaching the students whom the AI flags as struggling, while the AI handles the routine practice. Her students' grades have improved markedly. Her own working hours have decreased. She is, for the first time in her career, no longer dreading Sunday evenings.

What Jobs Will Actually Be Created?

The World Economic Forum's report lists, in considerable detail, the categories of work that will grow fastest by 2030 (WEF). Let me give you a flavour.

- AI and machine learning specialists.
- Data analysts and data engineers.
- Information security analysts, because AI also enables more sophisticated attacks.
- Renewable energy engineers, because the AI economy demands enormous power.
- Sustainability specialists.
- Care economy workers. Nurses, eldercare workers, mental health professionals, social workers, because AI does not replace human warmth.
- AI ethics, governance, and compliance specialists.
- Creative AI hybrid roles. Designers, writers, marketers, and producers who use AI to multiply their output.
- Skilled trades. Electricians, plumbers, technicians, agricultural specialists. (Yes, really. The unsexy hands on jobs are some of the most AI resistant.)

- Teachers and trainers, particularly those who can adapt to AI augmented classrooms.

If you have a child, a niece, a nephew, a younger sibling, or a student, this list is worth pinning to a wall.

Chapter Summary

We have looked, in this chapter, at the most personal question in the AI revolution. Will your job survive? The honest answer is. It depends. The WEF projects 78 million net new jobs globally by 2030, but the transition will displace 92 million workers in routine cognitive roles, and many of them will not seamlessly move into the new ones. We have identified four kinds of worker. The Resister, the Tourist, the Power User, and the Architect. And asked you which one you are. We have heard three stories of people who, by leaning into AI rather than away from it, transformed their careers. And we have looked at the categories of jobs most likely to grow in the decade ahead.

The job is not dying. The job, as you knew it, is dying. The new job is being born. You have the power to choose your place in it.

Coming Next

In Chapter Five, we turn to my own house, the university, and we look at an institution that has stood for almost a thousand years and is now, frankly, on fire. We will discuss what every vice chancellor, principal, dean, professor, parent, and student needs to know about the AI assault on higher education. Some of what I will say in the next chapter will not make me popular among my colleagues. But the time for polite silence is over.

❖ ❖ ❖

CHAPTER 5

THE UNIVERSITY IS ON FIRE

Saving Education From Extinction

In this chapter, we examine the institution closest to my own heart, the university, and we confront an uncomfortable truth. Its 900 year old model is in mortal danger. You will see what AI is already doing to teaching, assessment, research, and the very value of a degree. You will hear what I, as a vice chancellor, believe must change. And you will leave with a roadmap that schools, ministries of education, parents, and students can use to make sure the next generation is educated for the world that is coming, not the world that is passing.

I am about to say things in this chapter that my fellow vice chancellors will not enjoy reading. I have decided to say them anyway, because polite silence in a moment of historical disruption is a form of professional cowardice.

The university, as a model, was invented in Bologna in 1088. It has changed remarkably little in nine hundred years. A professor stands at the front. Students sit in rows. Knowledge is transmitted through lectures. Comprehension is verified through examinations. Successful students receive a piece of paper that signals to employers and to society that this person has been certified as knowing certain things.

Every single element of that model is now under assault.

The Three Pillars That Are Crumbling

The university traditionally rests on three pillars. **Teaching, assessment, and credentialing.** Let me show you how AI is shaking each of them.

Pillar One. Teaching. For nine centuries, the lecture was the gold standard of knowledge transmission. A professor who had spent decades mastering a subject stood before students and shared what she knew. The model worked because nothing else scaled.

Today, a 16 year old in rural Uganda with a smartphone and an AI tutor has access to one on one personalised instruction that no human professor on earth can match. The AI is patient. The AI explains things ten times if needed. The AI never tires, never has a bad day, never humiliates a slow learner. It can teach in Luganda, in Swahili, in English. It can adapt its difficulty in real time. It can generate infinite practice problems. And it costs, in many cases, less than a cup of tea per month.

The Adobe and Advanis "Creativity with AI in Education 2025" report surveyed 2,801 educators and found that **91 percent reported enhanced learning outcomes** when generative AI was paired with creative teaching practices (via Nucamp Uganda). A study of rural primary science classrooms in Uganda found AI tools boosted motivation, collaboration, and conceptual understanding (referenced in the same Nucamp source).

The lecture is not dead. But the lecture as the primary engine of teaching is finished. Whatever the university of 2035 looks like, it will not be primarily a place where one human stands at the front and twenty or two hundred listen.

Pillar Two. Assessment. For centuries, we have evaluated students by asking them to write essays, solve problems, and sit timed examinations. We have assumed that what a student produces is what a student knows.

In 2026, that assumption is shattered. A student today can submit an essay, a research paper, a code project, a marketing plan, a legal brief, a financial model, or an architectural sketch, and you, the marker, often cannot tell with certainty whether it was produced by the student, by the student's AI, or by some collaboration of the two.

Most universities have responded with two strategies, and both are failing.

Strategy one. Ban AI. This is futile. Students will use it anyway, in secret, and the gap between what we test and what they actually do in their careers will widen further. We will produce graduates who are AI illiterate, sent into workplaces that demand AI fluency. We will be training violinists in a world that needs pianists.

Strategy two. Deploy AI detection software. This is also failing. The detection tools are unreliable, frequently flag innocent students, and are always one step behind the generation tools. Worse, they create an adversarial atmosphere in which honest students live in fear and dishonest ones simply use better prompts.

What is the right answer? **Redesign assessment from the ground up.** Move from "what can you produce alone?" to "what can you produce *with* AI, and what is your particular contribution?" Move from take home essays to oral defences. Move from regurgitation to demonstration. Move from individual outputs to portfolio work that shows process, judgement, and iteration. Move from "can you find the answer?", a question AI now answers better than any of us, to "can you ask the right question, evaluate the answer critically, and apply it ethically in the real world?"

This is uncomfortable. It requires more work from professors. It requires new skills from administrators. It requires courage from faculty senates. And there is no alternative.

Pillar Three. Credentialing. Why does a university degree matter? Historically, because it signalled to employers that you had mastered certain knowledge, knowledge that was scarce and verified.

But knowledge is no longer scarce. Knowledge sits in everyone's pocket. What is scarce now is **judgement, creativity, ethics, collaboration, and the ability to integrate AI tools into meaningful real world output**. The university that continues to certify "she memorised the textbook" is certifying the wrong thing.

The Ellucian 2025 survey found that institutional adoption

of AI in higher education surged from 49 percent in 2024 to 66 percent in 2025, with 88 percent of administrators expecting it to keep rising (Ellucian). 43 percent of institutions now include AI in their strategic plans (Ellucian). The conversation has, at least, started. But conversation is not transformation.

What the Research Actually Shows About AI Tutoring

I want to pause here because, in every gathering of African vice chancellors and ministers of education I attend, someone always raises a familiar objection. "Lawrence, we cannot allow this in our classrooms. It will make our students lazy. It will cheat them of real learning." The objection is honest. The objection is also, on the evidence, wrong.

The research on AI tutoring has matured rapidly in 2024 and 2025, and the findings are now consistent across studies in three different continents. AI tutors, deployed well, raise learning outcomes meaningfully. AI tutors, deployed badly, do nothing or worse. Everything depends on the design.

A randomised study of seventh graders in the United States, led by researchers at Carnegie Mellon's Heinz College in 2025, found that students paired with a human tutor who was assisted by an AI tutor outperformed both unassisted human tutoring and unassisted AI tutoring on standardised mathematics gains (Carnegie Mellon, Heinz College). The point is not that AI replaces the teacher. The point is that the teacher plus the AI is a better teacher than either alone. This is the formula every education ministry on earth should be inscribing on the walls of its policy units.

A separate field experiment, published as an IZA discussion

paper in 2025, reported that an AI tutoring intervention in secondary school settings raised student learning measurably and, critically, did not crowd out the role of the human teacher. Students taught using AI tutors performed as well or better than peers, and reported feeling more comfortable asking questions of the AI than of a human in front of classmates (IZA Working Paper, AI Tutoring). The Brookings Institution, summarising the broader literature in 2024, concluded that generative AI tutoring is among the most promising innovations in education research in the last twenty years, with the largest gains observed for students who had previously fallen behind (Brookings).

Why does it work? Because of the simple, brutal economic truth that no education system on earth can afford a personal tutor for every child, and AI can. Bloom's two sigma problem, the famous educational research finding that students taught one to one outperform classroom students by two standard deviations, has tortured education economists for forty years. The solution was always too expensive. AI quietly, without fanfare, has made it affordable.

A child in rural Karamoja, sitting under a tree with a five year old phone and a free chatbot, can now have a patient tutor that explains photosynthesis as many times as she needs, in Luganda or Swahili or English, without ever growing tired or impatient. That tutor will not, by itself, transform her education. She still needs a human teacher who loves her, schools that are safe, parents who feed her, a national curriculum that respects her. But the marginal benefit of a free, patient, infinitely available tutor is the largest single educational technology gain since the textbook.

The institutions and countries that wake up to this fact in 2026 will, by 2030, be producing graduates the world wants to hire. The institutions and countries that ban these tools out of fear, or that adopt them carelessly, will produce graduates the world will quietly route around.

I lead Victoria University with this finding tattooed on my forearm. Not literally, of course. But morally, yes.

What the African Education System Must Do

Permit me, briefly, to speak directly to my own continent.

Africa has 75 percent of its population under the age of 30 in many countries (Nucamp Uganda). We have a demographic dividend the rest of the world envies. We also, in places, have school computer labs where fewer than 10 percent of installed machines function, and rural schools where 90 percent have no electricity (Nucamp Uganda).

We are sitting on the largest pool of young human potential in the world, and we are educating it for an economy that no longer exists.

This is the moment. Let me be blunt. If Africa misses this transition, we will spend the rest of the twenty first century as renters in a digital economy owned by others. If we seize it, we may yet write the most surprising chapter of global development that anyone has lived through.

Here is what I believe African ministries of education and universities must do in the next 24 months.

1. **Make AI literacy compulsory at every level.** Primary, secondary, tertiary. Not as a separate "computing" subject, but woven into every subject. Every history class teaches how to use

AI to research history. Every biology class teaches how to use AI as a study partner while critically evaluating its errors.

2. **Re train teachers, not replace them.** The most precious resource in African education is the teacher. AI does not replace her. It frees her from the mechanical drudgery of marking, lesson planning, and individualised remediation, so she can do what only she can do. Mentor, model, inspire.

3. **Invest in connectivity and power.** Uganda's government has set a connectivity target of 80 percent by 2030 (Nucamp Uganda). Every African country needs the same ambition. AI without bandwidth is a sports car without fuel.

4. **Build offline and low bandwidth AI tools.** Not every classroom will have 5G. Edge AI, models that run on cheap devices without the internet, is one of the most underrated opportunities for the global south. A bilingual AI tutor that works on a cheap phone with no signal could leapfrog an entire generation of educational infrastructure.

5. **Reform examinations.** Our national exam systems, Uganda Certificate of Education, West African Senior Secondary Certificate Examination, Kenya Certificate of Secondary Education, the matric in South Africa, all of them, were designed for an age when memorising facts mattered. They are mass producing graduates with skills the economy does not need. Reform is overdue, urgent, and political.

6. **Invest in local content.** The most beautiful AI tutor in the world is useless if it can only teach the curriculum of Massachusetts. We need African educators, African content creators, and African universities producing AI training data and applications in our

own languages, with our own examples, and grounded in our own contexts.

Victoria University and the Path I Have Chosen

I will not pretend that Victoria University, the institution I lead, has solved any of this. We are wrestling with the same questions as everyone else, and we are no further along than our peers. But I will share, briefly, the direction we are taking, in case it is useful for other institutions.

We have begun to weave AI literacy into every undergraduate programme. We have begun to retrain our faculty, not in how to code AI, but in how to teach with it and assess around it. We have invested in bilingual student facing AI tutors. We are redesigning examinations to test the things AI cannot do alone. Argumentation, oral defence, ethical reasoning, and the ability to use AI well.

It is the slowest, hardest work I have ever led. And it is also the most important.

Chapter Summary

In this chapter, we have looked at the three crumbling pillars of the traditional university, teaching, assessment, and credentialing, and we have seen that AI is shaking each of them. We have seen the survey data showing that 91 percent of educators report enhanced learning with AI, but also that institutional change is lagging. We have looked at what Africa must do in the next 24 months. We have heard, plainly, that banning AI in education is a losing strategy, and that the real work is redesigning what we teach, how we test, and what we certify. Education is the institution that, if it falls, brings everything else down with it. The good news is that AI may also be the technology that saves it, if we have the courage to let it.

Coming Next

In Chapter Six, we turn to the corner office. Every CEO reading this book, and every minister who runs a state owned enterprise, must wrestle with a single question. What does AI mean for my organisation? We will examine the most common mistakes business leaders are making, the patterns of those few who are getting it right, and the playbook for the eighteen months that will define whether your company is a champion of the AI economy or a footnote in its history. The next chapter is, perhaps, the most expensive one you will ever read. Measured in what it could save you, or cost you.

CHAPTER 6

THE CEO WHO REFUSED TO LOOK UP

A Warning to Business Leaders

In this chapter, we turn the spotlight on CEOs and senior leaders. You will see why the gap between AI ambition and AI execution is the defining business problem of this decade. You will read the survey data that should keep every board chair awake at night. You will meet the patterns of companies that are winning and the patterns of those quietly dying. And you will leave with a sober set of questions to ask yourself, your CFO, your CTO, and your team on Monday morning.

There is a moment in the film *Don't Look Up* when a comet is approaching the earth and the world's leaders cannot quite be bothered to take it seriously. It is a satire. It is also, in 2026, a documentary about how most of the world's CEOs are handling AI.

The data is brutal.

PwC's CEO Survey, conducted across hundreds of senior executives in over fifty countries, found that **fewer than a quarter of CEOs say AI is applied extensively across major business areas**, and most report limited financial returns to date (PwC CEO Survey). The Your AI Department benchmark of 164 CEOs found **40 percent are still in exploration or stalled pilot mode**, and only 27 percent have reached operational maturity (Your AI Dept). McKinsey reports that while 91 percent of higher ed administrators use AI personally, the gap between personal use and institutional integration is enormous, and the same pattern holds across most industries (Ellucian McKinsey context).

In other words. The CEO has played with ChatGPT on his iPhone in the back of his car between meetings. He has not yet figured out what to do with it in his actual company.

This is the gap that will make or break the next decade.

The Five Mistakes I See Almost Every Week

I sit on boards. I advise companies. I have one on one conversations with founders, CEOs, and ministers. Let me share with you the five mistakes I see repeated almost every week.

Mistake One. Treating AI as an IT project. The most common pattern. The CEO says, "We need an AI strategy," and immediately delegates it to the head of IT. The head of IT, an honourable and competent professional, then spends six months evaluating vendors, building a small pilot in a corner of the organisation, and producing a PowerPoint deck. Nothing changes in the day to day work of the company. AI is not an IT project. It is a business transformation project that happens to involve technology. The lead must be the CEO or a direct deputy with operational authority. Anyone less senior cannot move the organisation.

Mistake Two. Buying tools instead of redesigning workflows. The second most common pattern. The CEO subscribes to ChatGPT Enterprise, Copilot for Microsoft 365, and one or two specialised platforms, and considers the job done. But buying the tool is the easy part. The hard part, the part 90 percent of organisations skip, is redesigning *how the work flows* so the tool actually delivers value. McKinsey is explicit about this. The biggest gains come when AI is "embedded across entire processes," not deployed in isolated tasks (McKinsey).

Mistake Three. Underinvesting in change management. AI tools fail in organisations not because the technology is bad, but because employees do not use them, or use them wrong. A McKinsey study suggests that the largest determinant of AI success is not the model. It is whether the workforce adopts it. This requires training, incentives, role redesign, and patient leadership. Most CEOs underestimate this by an order of magnitude.

Mistake Four. Ignoring data quality. AI is only as good as the data you feed it. If your organisation's data lives in seventeen incompatible spreadsheets, three abandoned databases, and one heroic employee's head, no AI tool in the world will save you. The unsexy work of data cleanup is the precondition for the sexy work of AI deployment. Smart CEOs are doing this *now*, before they need it.

Mistake Five. Waiting for certainty. Perhaps the most expensive mistake. CEOs are trained to be careful with capital. They want a clear ROI before they commit. But AI is moving so quickly that the cost of waiting for certainty is now larger than the cost of acting under uncertainty. The competitor who launches in 2026 with a 60 percent certain plan will be miles ahead of the competitor who launches in 2028 with a 95 percent certain plan. PwC's data is clear. **Trust is now the number one blocker, ranked first by 26 percent of CEOs and in the top three by 46 percent** (Your AI Dept). And only 27 percent believe their leadership teams can even anticipate the disruption (PwC).

The Pattern of Companies That Are Winning

Let me give you the opposite. The pattern I see in the small minority of companies that are pulling ahead.

They appoint a senior AI leader who reports to the CEO. Not the CIO. A dedicated Chief AI Officer or VP of AI, with the authority to redesign workflows across departments. This person is part technologist, part change manager, part business strategist.

They start with one big bet, not fifty pilots. Pilot purgatory is real. The winners pick one core business process. Customer support, sales, recruiting, supply chain, marketing. And they reinvent it end to end with AI. Then they move to the next.

They invest in their workforce. They do not lay off staff when they deploy AI. They retrain them, redeploy them, and make the productivity gains visible to everyone. This creates allies for the next AI initiative, instead of enemies.

They have a data foundation. They cleaned up their data

before they needed to. Their AI projects work because the inputs are clean.

They tell stories. They publicise the wins, internally and externally. They make AI champions visible. They create a culture in which using AI well is celebrated rather than viewed with suspicion.

They have the CEO on the front line. The CEO uses AI. Personally. Daily. She drafts her own letters with it, summarises her own reports with it, prepares for her own meetings with it. She does not delegate this to her assistant. She models the behaviour she expects of her staff.

The Klarna Story Revisited. A CEO's Playbook

Let me return to the Klarna case from a different angle. Sebastian Siemiatkowski, the CEO, did several things right that are worth pulling out as a playbook.

First, he made AI a CEO level priority, not an IT level one. He personally championed the partnership with OpenAI and spoke publicly about it. Second, he redesigned a single core process, customer support, end to end, rather than scattering AI across the organisation. Third, he was transparent about the workforce implications, telling regulators, employees, and the public the truth. AI was doing the work of 700 agents. Fourth, when the model showed weaknesses on complex cases, he adapted. Klarna later announced it was rehiring human agents to handle the high judgement work (CBS News, LinkedIn coverage).

He moved fast. He moved publicly. He moved imperfectly. He moved.

That is the lesson.

The Hardest Conversation You Will Have

If you are a CEO reading this, the hardest conversation in front of you is with your board.

You will need to ask them to support investment in AI that will not pay off in the next quarter. You will need to ask them to fund retraining programmes that will not show up neatly in the P&L. You will need to ask them to tolerate experimentation, including failures. You will need to ask them to redefine the metrics by which they judge you.

Most boards, in 2026, are not yet wired for this. Their training, their incentives, their habits of mind, are pre AI. Many of them will resist, slow you down, or simply not understand what you are saying.

You have two choices. Change their minds, or, eventually, change your board. The CEO who cannot get her board on side will, in this decade, lose either her job or her company.

Chapter Summary

In this chapter, we have looked at the corner office honestly. We have seen that the great majority of CEOs are still in exploration or early pilot mode, that fewer than a quarter believe AI is applied extensively in their organisations, and that only 27 percent trust their leadership teams to anticipate disruption. We have catalogued the five most common mistakes. Treating AI as an IT project, buying tools without redesigning workflows, underinvesting in change management, ignoring data quality, and waiting for certainty. We have walked through the pattern of companies that are winning. And we have looked at the playbook implicit in the Klarna case.

The CEOs who will lead in 2030 are making different choices, today, than the CEOs who will be remembered as the ones who let the comet hit.

Coming Next

In Chapter Seven, we move from boardrooms to parliaments. We will look at the dilemma facing politicians and policymakers. How to govern a technology whose own creators do not fully understand its trajectory. We will examine the regulatory experiments of the European Union, the United States, China, Singapore, Rwanda, and the African Union. And we will discuss the duty politicians owe to their citizens in a moment when bad regulation could cripple a country for a generation.

❖ ❖ ❖

CHAPTER 7

THE POLITICIAN'S DILEMMA

Governing What You Do Not Understand

In this chapter, we examine the impossible, and unavoidable, task facing the world's politicians. Regulating a technology that even its creators cannot fully predict. You will see the different approaches taken by major economies, why some are likely to lead and others to lag, and what every minister, MP, and senior civil servant needs to do in the next 24 months. If you are a citizen, you will learn what to demand of your representatives. If you are a politician, you may finally have a framework for action.

I want to be honest with my politician friends, and with the politicians who do not yet know me but are reading this anyway. I have enormous sympathy for your position.

You are being asked to regulate a technology whose top experts disagree with each other about basic questions. You are being lobbied by industry voices that want light touch regulation, and by civil society voices that want stringent regulation. You are being told that if you over regulate, you will strangle innovation and watch your country fall behind. You are being told that if you under regulate, you will preside over the corrosion of your labour market, the surveillance of your citizens, and the destabilisation of your democracy. You are operating with a fraction of the expertise that the technology companies have. And you are expected to legislate in a 12 month news cycle for a 20 year transformation.

This is genuinely hard. So before I criticise, let me acknowledge. There are no easy answers.

But there are unforgivable inactions. Let me describe them.

The Three Failures I Will Not Forgive

Failure One. Pretending it is not happening. I have sat in cabinet briefings, I will not say in which country, where AI did not appear on the agenda. Not as a footnote. Not as a sidebar. Politicians who are responsible for the future of their citizens and who have not personally engaged with AI, who have not used it, who do not know what their own ministries are doing about it, are not doing their jobs. This is not optional knowledge. It is, today, as fundamental as understanding what a school is, what a road is, what a hospital is.

Failure Two. Outsourcing the thinking to consultants. I have seen too many governments, including, regrettably, on my own continent, commission a two million dollar McKinsey or Deloitte AI strategy report, file it on a shelf, and move on. The report is usually good. The follow through is usually nothing. Consultants cannot govern. Civil servants must own the thinking. Ministers must be trained.

Failure Three. Confusing regulation with strategy. Many governments today are racing to pass AI laws while having no AI plan. Regulation without strategy is like building speed bumps on a road you have not yet decided whether to pave. Every country needs both. A strategy for what AI will do for the economy, and a regulatory framework for what AI will not be allowed to do to the citizen. Singapore has both. The European Union has rapidly developed both. The UAE has both. Rwanda is building both. Most countries have neither.

A Quick Tour of How the World Is Responding

Let me give you a brief tour, with appropriate humility, of how different parts of the world are responding.

The European Union has passed the AI Act, the world's first comprehensive AI regulation, which classifies AI systems by risk level and imposes graduated obligations. It is detailed, it is ambitious, and it is widely admired and widely criticised in equal measure. Critics worry it will slow European AI development. Defenders argue it will give Europe a "Brussels effect" of setting global standards.

The United States has, characteristically, taken a more market led approach, with executive orders, voluntary commitments from major AI labs, and sector specific regulation. American AI development is the most aggressive in the world, but so is the American risk that regulation will eventually be reactive rather than proactive.

China has moved aggressively on both fronts. Building the second largest AI economy in the world, while imposing strict content controls and registration requirements. The Chinese state's relationship to AI is closer than any Western government's, with both the upsides, of coordinated industrial policy, and downsides, of privacy and dissent concerns, that implies.

Singapore quietly created the first comprehensive national AI strategy a decade ago and has continued to update it. It is, per capita, perhaps the best prepared country in the world.

The United Arab Emirates appointed the world's first Minister of AI in 2017, His Excellency Omar Sultan Al Olama, and has since become a regional AI hub. This is what political seriousness looks like.

Rwanda has, under President Kagame, become Africa's AI policy leader, with a national AI policy and aggressive partnerships with global firms.

Uganda, my own country, has begun, with a National AI Task Force, a partnership with UNESCO for an AI Readiness Assessment, and a developing regulatory framework that emphasises human rights (Nucamp Uganda). I want to commend the work being done. I also want to say, with respect, that it must accelerate. Twelve month timelines are too slow. The technology will not wait.

The Seven Things Every Government Must Do Now

If I could write a single page for every cabinet on earth, it would read as follows.

One. Make every minister AI literate. Not optional. Mandatory training. Personal use of the tools. Demonstrated competence. A minister who has never used ChatGPT in 2026 has no business making decisions about AI in any portfolio.

Two. Appoint a senior AI lead in every ministry. Not just in technology or innovation. Health, education, agriculture, finance, defence, justice. Every ministry needs a senior officer whose job is to understand and deploy AI in that portfolio.

Three. Publish a national AI strategy. With targets. With timelines. With budgets. A strategy without numbers is a wish.

Four. Pass a balanced AI law. Risk based, sector aware, internationally aligned. Protect citizens. Do not strangle innovation. Update it annually.

Five. Invest in national AI infrastructure. Connectivity. Power. Compute. Data infrastructure. Local language models for low resource languages. Public AI compute capacity for universities and startups.

Six. Massively scale workforce reskilling. Every government needs an AI era workforce strategy as serious as its school strategy. Hundreds of thousands of citizens need new skills.

Seven. Lead by example. Government services should be among the first to use AI well, for citizen services, healthcare delivery, agricultural extension, public health communication. When citizens see government using AI to serve them better, trust grows.

When they see government using it to surveil them, trust collapses. Choose well.

A Story I Hope to Tell One Day

I want, before I close this chapter, to share a small dream.

I dream of standing in a school in Karamoja, in the dry north of Uganda. A small girl is sitting at a battered desk with a cheap tablet. She is asking her bilingual AI tutor, in Karamojong, a question about her science homework. The AI is answering her, patiently, with examples drawn from the cattle and crops of her region. She is learning. She is laughing. She is, in that moment, every bit as well resourced educationally as a child in Oslo or Singapore.

That moment will not happen by accident. It will happen because, at some point in 2026 or 2027, a government chose to invest in it.

That is what politics is for.

Chapter Summary

In this chapter, we have looked at the politician's dilemma honestly. We have sympathised with the difficulty, and we have refused to forgive the inaction. We have catalogued three unforgivable failures. Pretending it is not happening, outsourcing the thinking, and confusing regulation with strategy. We have toured the responses of the EU, the US, China, Singapore, the UAE, Rwanda, and Uganda. And we have laid out seven concrete steps every government must take in the next 24 months. Politicians have a unique role in the AI transition. They are the only actors who can ensure the gains are shared, the harms are mitigated, and the slowest among us are not left behind.

Coming Next

In Chapter Eight, we will turn fully to my own continent. Africa stands at a once in a civilisation moment. We can either be the great success story of the AI age, or its greatest failure. I will share what I believe, with all my heart, must happen for it to be the former. This next chapter is, frankly, the one I have been waiting my whole life to write.

❖ ❖ ❖

CHAPTER 8

AFRICA'S ONCE IN A CIVILISATION CHANCE

Why the Coming Decade Will Decide Our Place in the Twenty First Century

In this chapter, I make the case for Africa. You will see why our youth dividend, our mobile first economies, our untapped languages, and our hunger for change make us, paradoxically, one of the best positioned regions in the world to leapfrog in the AI age. You will also hear, with characteristic frankness, the traps we must avoid. If you are African, this chapter is a love letter and a warning. If you are not African, it is an invitation to bet on us, because the smart money increasingly is.

I have written most of this book in a tone meant to address everyone, from a CEO in Berlin to a teacher in Bangalore. In this chapter, with the reader's permission, I will speak more directly from where I sit. I am African. I am a son of Uganda. I lead a university in Kampala. And I believe, with every fibre of my being, that the next twenty years can either be Africa's greatest century, or its most painful one. And I believe AI will be the deciding factor.

What We Have Going for Us

Let me start with the good news, because it is genuinely good and often underappreciated.

Demographics. Africa is young. Insanely young. Median age in many African countries is under 20. In Uganda, more than 75 percent of the population is under 30 (Nucamp Uganda). In a world where most economies are ageing rapidly, Japan, Germany, Italy, China, Africa is the great pool of young human energy. Young brains absorb new technology faster. Young workforces are more adaptable. Young populations are more entrepreneurial.

Mobile first economy. Africa skipped the desktop computer. We went straight to mobile. M Pesa transformed Kenya into one of the most mobile money developed economies on earth. WhatsApp is already the operating system of African commerce. This means that the AI revolution, which is increasingly mobile first, is built on infrastructure we already use better than many developed countries.

Hunger. This is the most important factor and the one I cannot overstate. Africans are hungry. Our young people are hungry for skills, opportunity, mobility. There is no satiated, comfortable, change averse middle class blocking innovation the way you sometimes see in other regions. We will run faster, when given a chance, because we have further to run.

Languages. Africa has over 2,000 languages. This is sometimes painted as a barrier to AI development. I see it as the opposite. It is a massive, untapped data frontier. The teams that build the first truly fluent Luganda, Swahili, Yoruba, Amharic, isiZulu, Hausa, and Wolof AI models will create enormous economic and cultural value. This is a global opportunity sitting in our laps.

Real problems waiting to be solved. Western AI applications are often solving first world conveniences. Slightly better email drafting, slightly better movie recommendations. African AI applications can solve civilisation shaping problems. Malaria diagnosis in a village clinic, cassava disease detection on a small farm (Crazy Kanairo Farming), legal access for the unrepresented, agricultural advice where extension officers do not exist. The impact per dollar of African AI applications is enormous.

What We Have Working Against Us

Let me also be honest about our challenges, because pretending they do not exist serves no one.

Infrastructure. Connectivity remains inadequate in much of rural Africa. In Uganda alone, more than a thousand school computer labs are installed but fewer than 10 percent are functioning, and up to 90 percent of rural schools have no electricity (Nucamp Uganda). You cannot have an AI economy on a 2G connection in a school with no power. This must be the first priority.

Skills. We do not have nearly enough AI engineers, data

scientists, AI product managers, AI ethics specialists, or AI trained educators. Filling this pipeline is the work of a decade.

Capital. The vast majority of global AI capital flows to the United States and China. African startups in AI raise a small fraction of what their global peers raise. This must change, and it will only change when the rest of the world sees a returns story here.

Governance. Some of our governments are leading well. Some are not engaging. A few are actively destructive. AI requires a baseline of institutional capacity that not every state currently provides.

Brain drain. Our brightest young people are increasingly being pulled to Berlin, London, Toronto, San Francisco. Many will not come back. This is heartbreaking and it is a fact. We must build careers, institutions, and salaries at home that compete.

The Three Leapfrogs Africa Must Make

I believe Africa has the opportunity to make three civilisational leapfrogs in the next decade, each enabled by AI.

Leapfrog One. From low productivity to high productivity agriculture. AI powered tools for soil analysis, crop disease detection, weather prediction, and market access can transform smallholder farming. Kenyan farmers in 2026 are already using offline capable AI apps like PlantVillage Nuru, Tumaini, and Agrio to fight pests and diseases on cassava, maize, and bananas (Crazy Kanairo Farming). Farmer Lifeline Technologies offers AI powered devices that continuously scan farms and send real time alerts (Farmer Lifeline). Scaled across the continent, this is the equivalent of a green revolution, but cheaper, faster, and more equitable than the one of the 1960s.

Leapfrog Two. From under served healthcare to AI augmented universal access. A village in northern Uganda may never have a resident specialist. But it could have a smartphone with an AI diagnostic assistant, telehealth links to specialists in Kampala or Nairobi, and AI powered portable X ray devices (Nucamp Uganda). The clinical accuracy gains are already documented. AI in radiology improves sensitivity by 8 percent and reduces reading time by 52 percent (PMC NCBI), and these gains can be deployed in any clinic with bandwidth.

Leapfrog Three. From under resourced education to personalised, bilingual, AI augmented learning. This is the one closest to my heart. A bilingual AI tutor in Luganda or Swahili, running on a cheap device, available 24 hours a day, can give a child in Karamoja or Kakuma access to a quality of personalised instruction that a child in Manhattan cannot afford. We can leapfrog the resource constraint that has bound us for generations.

These leapfrogs are not science fiction. The technology exists. The proof points exist. What is missing is leadership, capital, and execution.

What Every African Leader Must Do

To my fellow Africans in positions of leadership, vice chancellors, ministers, CEOs, principals, mayors, religious leaders, traditional leaders, let me make this as concrete as possible.

Take an hour today and use an AI tool yourself. Not next week. Today.

Hold a meeting this month, with your team, your cabinet, your faculty, your board, about your AI strategy. If you do not have one, start drafting.

Pick one process in your organisation and reinvent it with AI in the next quarter.

Invest in training your people. Not in lavish overseas conferences. In practical, hands on training in the tools that exist today.

Tell stories of African AI excellence. Promote the entrepreneurs, the academics, the students who are building. Make heroes of them. Africa's AI culture will be built on visible role models.

Demand of your government what your government should already be demanding of itself. Connectivity, power, education reform, regulatory clarity, capital.

The Money Africa Can Capture, If We Act Now

Let me share the numbers that should be circulating in every African cabinet meeting, every parliamentary committee, every university council, and every chamber of commerce.

SAP's research, published in mid 2025, projects that AI will add roughly 1.5 trillion United States dollars to Africa's GDP by 2030, provided we adopt aggressively and invest in skills and infrastructure (SAP News, Africa AI 2030). A separate expert panel reported by BusinessDay in 2026 estimated that AI could lift Africa's combined GDP by 2 trillion dollars by 2035 (BusinessDay, Africa AI 2035). The United Nations University's Institute for Natural Resources in Africa has published a parallel analysis concluding that AI is one of the most plausible accelerants for African economic transformation in the coming decade (UNU INRA).

One and a half trillion dollars. Two trillion dollars. These are sums larger than the entire combined annual budget of every African government today. They are sums larger than every official development assistance flow we have ever received. They are sums that would, if captured, allow this continent to finance its own roads, its own hospitals, its own universities, its own scientific research, without lining up outside the doors of foreign capitals.

The condition is the same as everywhere else. We must act, and we must act now. Last week is too late. Next year is too late. Today is the day to begin.

The early signs are encouraging. East African governments, working through the East African Community, have begun coordinating an AI strategy with a regional fund, a sovereignty oriented framework, and a plan for a Ugandan AI factory that would manufacture, train, and serve African language AI models from African soil (iAfrica, East Africa AI). Uganda's New Vision has reported sustained debate in our parliament about the inclusion of the private sector in our national AI policy, with industry leaders calling for an enabling regulatory environment that recognises AI as a strategic national priority (New Vision Uganda).

Health ministries across the continent are beginning to pilot AI driven diagnostic systems, with research published by African health journals showing that AI assisted triage and imaging in resource constrained settings can dramatically improve outcomes (JSciMed Central, AI African Healthcare). Kenyan startups are closing data gaps in healthcare using AI, with Google reporting on one Nairobi based firm whose tools are now used across community health worker networks (Google Africa Blog).

None of this is enough yet. None of these initiatives, taken alone, will deliver the trillion dollar upside. But they are evidence that the continent is beginning to stir. The question is whether we will stir fast enough to seize the moment, or wake too late and watch others harvest the value of our young people, our data, our markets, and our problems.

I tell every African leader I meet, in my own country and beyond, that the AI window will not stay open forever. The infrastructure decisions made between 2026 and 2030 will shape African economic trajectories for the rest of the century. We get to choose. We must choose courageously.

Chapter Summary

In this chapter, I have made the case that Africa is, paradoxically, one of the best positioned regions in the world for the AI age, because of our demographics, our mobile first habits, our hunger, our languages, and the magnitude of the problems waiting to be solved. I have also been honest about our infrastructure, skills, capital, governance, and brain drain challenges. I have proposed three civilisational leapfrogs, in agriculture, healthcare, and education, that AI makes possible. I have shown that SAP and the United Nations University converge on a forecast of around 1.5 trillion dollars in additional African GDP by 2030, and that some panels see 2 trillion by 2035, provided we act. I have pointed to encouraging early signs. The East African AI coordination, the Ugandan AI factory plan, Kenyan healthcare startups, and the slow stirring of our policy systems. And I have asked African leaders to act, today, not in five years.

The next twenty years will decide whether the African century is the one we have been waiting for or another deferred dream. The pen is, for once, in our hand.

Coming Next

In Chapter Nine, we will get tactical. We will look at the single most undervalued idea in the AI age. That one well prepared human, equipped with AI, can now produce the output of ten humans without it. We will examine, with examples, how this works in practice for writers, marketers, lawyers, scientists, entrepreneurs, and educators. If you are a manager, a founder, or an ambitious individual, the next chapter is the one to dog ear and return to.

❖ ❖ ❖

CHAPTER 9

THE PRODUCTIVITY MULTIPLIER

How One Human Becomes Ten

In this chapter, we get practical. You will see, with concrete numbers and real examples, how AI is enabling individual humans to produce two, five, or even ten times their previous output. You will learn the basic disciplines of productive AI use, prompting, verifying, iterating, and integrating, that separate Tourists from Power Users. And you will see why this productivity multiplier, more than any other single factor, will determine who thrives in the next decade.

If there is one phrase you should take from this book and write on a sticky note above your desk, it is this. **AI is a productivity multiplier, not a productivity replacer.**

The companies that get this win. The individuals who get this win. The countries that get this win. The ones that do not, lose.

Let me show you what this looks like in numbers, then in stories, then in practice.

The Numbers Are Real

GitHub conducted a famous controlled experiment in which 95 developers were split into two groups. One group used GitHub Copilot, an AI coding assistant. The other did not. They were given the same task. Implement an HTTP server in JavaScript. The Copilot group completed the task **55.8 percent faster** than the control group. 1 hour 11 minutes versus 2 hours 41 minutes (GitHub Research, arXiv paper). This is not a marketing claim. It is a peer reviewed, controlled study.

In radiology, ProFound AI has been shown in clinical trials to enhance radiologist sensitivity by 8 percent while reducing reading time by 52 percent (PMC NCBI). The radiologist is not replaced. She is multiplied. She finds more cancers, faster, and gets to spend more time on the cases that require her judgement.

In customer service, Klarna's AI chatbot did the equivalent work of 700 full time agents while simultaneously reducing repeat enquiries by 25 percent (CBS News).

These are not isolated cases. Across knowledge work, productivity gains in the 30 to 70 percent range are now commonplace for the workers who genuinely integrate AI into their daily flow. McKinsey's analysis of long term corporate AI value sits at 4.4 trillion dollars in productivity potential (McKinsey).

A Day in the Life of a Power User

Let me paint a picture of what this productivity multiplier looks like in practice. This is composite, drawn from several real professionals I know, but every detail is true to life.

Sarah is a mid level marketing manager at a Nairobi based fintech. Her old day.

- 8:00 am. Catch up on overnight emails. 45 minutes.
- 8:45 am. Weekly team meeting. 1 hour.
- 9:45 am. Review and edit a draft press release. 2 hours.
- 11:45 am. Research a new market segment. 3 hours.
- 2:45 pm. Lunch and email.
- 3:30 pm. Write a strategy memo for next month's campaign. 2.5 hours.
- 6:00 pm. Leave the office, exhausted.

Sarah's new day, with AI integrated.

- 8:00 am. Open her email assistant, which has already summarised overnight emails into three categories. Respond now, respond later, ignore. She handles the "now" category in 15 minutes.
- 8:15 am. Spend 30 minutes deeply preparing for her 8:45 meeting with the help of AI that has summarised relevant docs.
- 8:45 am. Weekly meeting. She arrives sharper.
- 9:45 am. AI drafts the press release in 5 minutes. She spends 25 minutes refining it. Done in 30 minutes, not 2 hours.

- 10:15 am. Market research. She prompts the AI to compile a structured briefing on the new segment. She verifies the key numbers personally. Done in 1 hour, not 3.
- 11:15 am. She starts on the strategy memo. The AI helps her brainstorm, structure, draft. She does the high value thinking. It does the mechanical writing. Done by 12:30 pm.
- 12:30 pm. She has time for a proper lunch, a walk, and an afternoon she would not have had before.

She has done a day's work in half a day. She is not less valuable to her employer. She is much more valuable. She is producing strategic outputs, not just sweating through mechanical ones. And the employer, if smart, does not lay her off. The employer either gives her more strategic work, or rolls out more campaigns with her, or expands into new markets she could not previously have served.

This is the multiplier. This is what it looks like.

The Four Disciplines

Becoming a Power User is not magic. It is the disciplined practice of four habits.

Discipline One. Better prompting. AI gives back what you put in. The single biggest difference between Tourist and Power User is how clearly they ask for what they want. A Tourist types "write me an email." A Power User types. "Draft a 200 word follow up email to a senior client we last met three weeks ago, referencing their interest in supply chain financing, with a polite but specific suggestion for a 30 minute call next week. Tone. Warm, professional, not desperate." The Power User's email comes back ready to send. The Tourist's needs five revisions.

Discipline Two. Disciplined verification. AI hallucinates. AI gets numbers wrong. AI cites sources that do not exist. Every Power User has the iron habit of verifying any factual claim, any citation, any specific number, before relying on it. AI is a brilliant assistant, not a trusted source. Treat its outputs as a draft, not a fact.

Discipline Three. Iterative refinement. A Tourist asks once and accepts the answer. A Power User has a conversation. "That is good. Now make it shorter." "Now add a section on risk." "Now rewrite it for a non technical audience." The output that emerges after five iterations is dramatically better than the first attempt.

Discipline Four. Workflow integration. A Tourist uses AI in isolated moments. A Power User builds AI into their daily processes. Email triage, meeting prep, research, drafting, code, planning, summarising, learning. The compound effect is what produces the 10x outcome.

A Note on Honesty

I want to insert a short note here, because the productivity multiplier conversation often slips into hype.

Honesty matters. Power Users disclose, where appropriate, that they used AI. They do not pretend the AI's contribution was their own. They do not, in academic settings, submit AI work as personal work. They do not, in legal settings, file briefs with fabricated citations. The Power User uses the tool transparently and responsibly, and is therefore more trusted, not less.

The Power User is also honest about AI's limits. She tells her colleagues when AI got something wrong. She maintains her own expertise. She does not let her brain atrophy. She uses AI to be a better version of herself, not a hollow one.

Chapter Summary

In this chapter, we have looked at the most practical, immediate consequence of the AI revolution. The productivity multiplier. We have seen the data. 55.8 percent faster coding (GitHub Research). 52 percent faster radiology reading with 8 percent better cancer detection (PMC NCBI). 700 agents worth of customer service work done by software (CBS News). We have walked through a composite day in the life of a Power User. We have identified the four disciplines, better prompting, disciplined verification, iterative refinement, and workflow integration, that separate Tourists from Power Users. And we have noted the importance of honesty in using AI well.

If everything else in this book is theoretical to you, let this chapter be practical. Pick one workflow. Apply the four disciplines. Watch your output double. Then do it again.

Coming Next

In Chapter Ten, we will pause for a moment to slow down and translate. The most important word in the AI conversation of 2026, the word agentic, has been thrown about by journalists and consultants without anyone quite stopping to explain what it actually means in the life of an ordinary woman or man. I will fix that. I will introduce you, by name, to the new generation of digital apprentices already changing how millions of people work. Claude Cowork. OpenAI's Codex. And I will tell you stories of farmers, lawyers, course creators, and small business owners who are quietly becoming twice the workers they used to be, because they have learned to befriend these agents. After that, in Chapter Eleven, we will look squarely at the dangers.

❖ ❖ ❖

CHAPTER 10

THE APPRENTICE THAT NEVER SLEEPS

Agentic AI in Plain Language, and Why It Changes Everything

In this chapter, I want to take the most important and most misunderstood word in the AI conversation of 2026, the word agentic, and put it in language a market trader in Owino, a teacher in Mbarara, a banker in Kampala, and a grandmother in Jinja can all understand. I will tell you what an AI agent actually is, with stories, not jargon. I will introduce you to the new generation of these agents, Claude Cowork, OpenAI's Codex, and others, the way you would introduce a new employee. I will show you, through real examples, how ordinary people are using them to do in one afternoon what used to take a small team a whole week. By the end of this chapter, you will not just know the word agentic. You will know why everyone who does not befriend an agent in the next two years will be left several streets behind everyone who does.

Let me begin with a small confession.

For the first two years of my own AI journey, I used these tools the way most people still use them today. I would open a chat window, type a question, read the answer, copy and paste what I needed, and close the window. The machine helped me. It made me faster. It made me more thoughtful. But it was, in essence, a very smart library and a very patient typist. I had to ask it everything. I had to read its work. I had to take what it produced and walk it over to the next step myself, like a man carrying buckets of water from a well to a kitchen.

Then, sometime in the last twelve months, something changed.

The machine learned to walk to the well by itself.

I would describe the task once. The machine would think about it, plan the steps, open the right tools on my own laptop, read the files it needed, write the document, send it to the printer, schedule the follow up meeting on my calendar, and ping me only when it ran into a question that genuinely required my judgement. I was no longer the water carrier. I was the foreman of a small, tireless apprentice who never needed to eat, never grew bored, and never asked for a raise.

This shift has a name. Agentic AI.

And it is the most important shift in technology since the invention of the personal computer itself.

What an Agent Actually Is

Let us strip the word of all its mystery.

An agent, in the ordinary English sense of the word, is someone you send to do something on your behalf. A real estate agent. An insurance agent. An agent at the immigration desk. A football agent who negotiates with clubs while the footballer focuses on football.

An AI agent is the same idea, except the someone is a piece of software, and the something can be almost anything you can describe in plain language.

There are three differences between an AI agent and the chatbots most people are familiar with.

Difference number one. An agent can take action, not just give answers. A chatbot tells you the best flight from Entebbe to Dubai. An agent finds the flight, compares the prices across airlines, checks your calendar for clashes, makes the booking, pays from your card, adds the trip to your calendar, books your hotel in Dubai, and emails the confirmation to your assistant. All from a single instruction in plain English.

Difference number two. An agent can use tools. It can open your spreadsheet and edit it. It can write a document and save it in the correct folder. It can search the web, read the results, and decide which ones to trust. It can run software code. It can fill in a web form. It can read a PDF. It can send a message on Slack or WhatsApp. It can make a phone call through a voice service. It can, in short, do almost anything a junior staff member sitting at a laptop could do, except faster and at three in the morning.

Difference number three. An agent can plan. Give it a goal that requires twenty separate steps and it will work out the order of those steps, do them one by one, check its own progress, correct itself when it makes mistakes, and tell you what it has done when it is finished. It is not just reactive. It is, in a real and useful sense, purposeful.

Put these three differences together and what you have is no longer a chatbot. It is a digital employee. An apprentice who has read every book ever written, never gets tired, costs less than a cup of coffee per hour, and is improving at a rate no human can match.

Meet the New Generation of Apprentices

There are several of these apprentices on the market today. The two most talked about, the two I want you to know by name, are Claude Cowork from a company called Anthropic, and Codex from OpenAI, the company that makes ChatGPT.

The names do not matter as much as what they do. Think of them the way you think of Toyota and Mercedes, or Apple and Samsung. They are competitors building broadly the same thing, each with its own personality and strengths. What is important is that, as of 2026, both of them have moved beyond the chat window and into the realm of true agency.

Claude Cowork is built by Anthropic, the company many in the AI safety world consider the most careful of the major laboratories. The name says everything you need to know about the product. Cowork is what Claude becomes when it stops merely answering questions and starts showing up to work alongside you. It runs on your computer. It can see your screen. It can open your applications. It remembers the project across many sessions, the way a human colleague would remember what you discussed yesterday. It can be given a long running task and left alone for hours, while you sleep or attend a meeting, and you will return to find the work done (Build to Launch, Claude Cowork case studies).

OpenAI's Codex began life as a tool for software engineers. Today it has grown into something much wider. As of April 2026, OpenAI rebranded the entire product as Codex for almost everything, and gave it the ability to use your computer like a person would, with computer use on macOS, an in app browser, image generation, persistent memory, scheduled tasks, and more than ninety plug ins for tools like Jira, Microsoft 365, Notion, Slack, HubSpot, and Salesforce (Spicy Advisory, April 2026 Codex update). The marketing slogan is exactly right. Codex is now for almost everyone, not just programmers.

There are others. Google has Gemini's agentic mode. Microsoft has Copilot agents. Meta is working on its own. The competition is fierce, and you, the user, are the beneficiary.

But here is the point I want to make plainly. You do not need to know all the names. You need to know one truth. Every serious computer in 2027 will have an agent built into it, the same way every car in 1980 came with an engine. The question is not whether you will use one. The question is whether you will use one well.

The Story of Margot's Hour

Let me tell you a story I read recently, because stories teach better than definitions.

A woman named Margot, living in Houston, Texas, wanted to buy a used car. The kind of car she wanted was rare and the dealerships in her city often had only one or two examples in stock at any given time. The traditional approach to her problem would have taken weeks. She would have had to search dealership websites one by one, write down phone numbers, call during business hours, leave messages, wait for callbacks, repeat herself to each new salesman, take detailed notes, compare prices, schedule visits, and so on. A part time job in itself.

Instead, she gave the task to Claude Cowork.

Within one hour, the agent had built a tracking spreadsheet of every dealership in Houston that met her criteria. It had drafted personalised outreach emails to each one, filled in the web inquiry forms on their websites, and managed the reply threads. Within that same hour, several dealerships had responded with inventory details. Claude then handled the follow up questions in a second round of communication, while Margot did her actual day job (Build to Launch, Claude Cowork case studies).

She did not become a worse car buyer. She became a much better one. She got more options. She paid less. She wasted no time. And the work happened while she lived her life.

This is what agentic AI looks like in practice for an ordinary person. Not a robot uprising. Not a science fiction film. A small, polite assistant that handles the boring parts of life so you can spend your hours on the parts that matter.

The Story of Dheeraj's Course

A second story, because the diversity of these use cases is the point.

A man named Dheeraj was building an online course. Forty seven lessons. Many overlapping topics. The kind of project where, by lesson thirty, you have forgotten what you said in lesson seven, and you risk repeating yourself, contradicting yourself, or, worse, building a confused experience for the student.

He gave the entire course folder to Claude Cowork. Because the agent maintained persistent context across all the lessons, it caught overlaps. It cross referenced earlier lessons. It flagged contradictions. It suggested transitions. He told me, in his own words, "That was the moment I realised the project level context is what makes Cowork different from just chatting with Claude in a regular conversation" (Build to Launch, Claude Cowork case studies).

This is what agentic AI looks like in practice for a creator. A patient editor who has read every word you have ever written for this project and remembers it as clearly as if you wrote it ten minutes ago.

The Story of the Seven Hour Audit

A third story, this time from the world of large enterprise.

The Japanese e commerce giant Rakuten ran a test. They took a codebase of twelve and a half million lines of computer code, the kind of mountain only the largest companies in the world maintain, and they let Claude Code, the engineering twin of Cowork, work on it autonomously for seven hours. A single, unsupervised run.

The result. Ninety nine point nine percent accuracy (Anthropic 2026 Agentic Coding Trends Report).

To put that in context. A team of ten skilled engineers might take a month of careful work to perform the equivalent audit. The agent did it in an afternoon. Without a single coffee break. Without a single complaint about overtime.

This is what agentic AI looks like at the frontier. Not magic. Not science fiction. Just an apprentice with patience no human possesses, running at the speed of electricity, finishing what would have taken a team weeks before lunch was served.

The Anthropic Legal Team and the Twenty Four Hour Window

The fourth story is one of my favourites because it disproves the most common myth about agentic AI.

The myth says that you must be a programmer to benefit from these tools. The myth is wrong, and the people who are quietly disproving it are not engineers at all.

Anthropic's own internal legal team, the lawyers who handle contracts, compliance, and reviews for one of the most important AI companies in the world, decided to build their own tools using Claude Code. Not one of them was a software engineer by training. They were lawyers. They were people who, like most of you reading this, had spent their careers reading documents, writing memos, and arguing in meetings.

Within a few months, those lawyers, using nothing but plain English instructions, had built internal tools that reduced their contract review time from two to three days down to twenty four hours (Anthropic 2026 Agentic Coding Trends Report).

Read that sentence again. Lawyers, with no coding background, built tools that cut their workload by sixty to seventy percent. Across a busy legal department, this is the difference between hiring three more associates and not having to.

The same report notes that the company Zapier achieved eighty nine percent AI adoption across its entire organisation, with more than eight hundred AI agents deployed internally to handle everything from design prototyping to customer interviews (Anthropic 2026 Agentic Coding Trends Report).

Agentic AI is not for programmers. It is for everyone who has a job that involves a computer. Which, in 2026, is almost everyone.

A Day in the Life of a Person Who Has Befriended an Agent

Let me try to paint, in simple strokes, what an ordinary professional day looks like for someone who has truly embraced agentic AI in 2026. The picture is not science fiction. It is what is happening, right now, in offices from Lagos to Boston.

Imagine a woman called Sarah. She is the head of programmes at a small NGO in Kampala. She has seventeen direct reports, four funders, and more emails than hours in the day.

Six thirty in the morning. She opens her phone. Her agent has already, while she slept, gone through her overnight emails, summarised the urgent ones, drafted replies to the routine ones for her approval, flagged three that require her personal attention, and prepared a one paragraph briefing on the donor meeting she has at ten.

Seven thirty. She drives her children to school. While she is in the car, her agent is on a scheduled task. It is pulling the monthly numbers from her project management system, building a draft of the quarterly report, and producing the first version of the budget reconciliation. By the time she sits at her desk at half past eight, the report is waiting, ready for her to review and refine.

Nine. She begins reviewing. The work is not perfect. There are three places where the agent has misunderstood the context, and she corrects them. There is one paragraph that needs her personal voice, and she rewrites it. There is a chart that needs to be redrawn, and she instructs the agent to redraw it, which it does in thirty seconds.

Ten. Donor meeting. Because her agent has been listening, with permission, throughout the call and taking notes, she can focus entirely on the conversation rather than scribbling. When the meeting ends, the agent has a full transcript, a summary of action items, and draft follow up emails ready for her approval.

Eleven. She eats a samosa and drinks tea, because she still has time to.

Eleven thirty. A team member submits a project proposal. Sarah forwards it to her agent and asks for a review against the funder's stated criteria. Five minutes later she has a detailed analysis showing what is strong, what is weak, what is missing, and how the proposal compares to three similar proposals the team has submitted in the past two years.

Two in the afternoon. A staff conflict surfaces. This is not a task for the agent. This is a task for a human leader, and Sarah handles it herself. The agent's job is to handle the routine so that Sarah has the patience, the time, and the presence of mind to handle the human.

Five in the evening. She closes her laptop. The day is done. The work that would have kept her until ten at night is already finished. She drives home. She has dinner with her children. She reads. She sleeps.

This is what agentic AI does for a person who learns to befriend it. It does not replace her. It does not threaten her. It gives her, back, the most precious commodity a working parent has. Time.

The Pattern That Repeats Across Every Industry

Now, look at that day in Sarah's life and substitute different professions. The pattern is the same.

A farmer in Mbale sends a photograph of his cassava leaves to an agent on his phone. The agent diagnoses the disease, recommends the treatment, locates the nearest supplier, places the order, and arranges delivery. While the farmer is in the field.

A teacher in a rural school asks her agent to grade a stack of fifty essays against a rubric. The agent produces individual feedback for each student, flags the three essays that need the teacher's personal attention because they touch on emotional issues she should not delegate, and prepares a one page summary of the class's strengths and weaknesses. The teacher spends her evening doing what only she can do, which is talking to the three children who need to be talked to.

A nurse in a rural health centre uses an agent to triage patients.

The agent listens, in the patient's own language, to the description of symptoms, asks the right follow up questions, suggests the most likely conditions, recommends the appropriate next step, and alerts the doctor only when the case is complex or urgent. The nurse handles four times as many patients as before, with better outcomes.

A small business owner in Owino market asks her agent to manage her social media. The agent posts three times a day, in the right tone, with the right images, in three languages. It replies to comments. It captures interested customers and routes them to her WhatsApp. The shop owner does not have an MBA. She does not have a marketing team. She has an agent and the wisdom to use it.

A pastor in Kabale uses an agent to research his sermon. He gives the agent his theme, his preferred translation of scripture, and a few illustrations from his own life. The agent returns with a structured outline, three relevant historical examples, the text of two old hymns that fit the theme, and a one paragraph summary suitable for the church bulletin. The pastor's sermon is better. His preparation time has dropped from a full day to two hours. The two hours he saved are the two hours he now spends visiting a sick member of his congregation.

In every one of these stories, the agent is not the hero. The human is the hero. The agent is the apprentice. But what an apprentice.

What the Honest Data Says About Agentic AI

I want to ground all this storytelling in numbers, because I want you to know that these are not anecdotes. They are the leading edge of a tidal shift.

Anthropic's 2026 Agentic Coding Trends Report, an eighteen page study of how the world's most advanced developers are using these tools, found that developers now use AI for approximately sixty percent of their work, although they only fully delegate between zero and twenty percent of their tasks (Anthropic 2026 Agentic Coding Trends Report). What does this mean. It means agentic AI is not replacing the human. It is augmenting the human, deeply.

Twenty seven percent of all AI assisted tasks, the report found, are tasks that would never have been done at all without AI. Internal tools that no one had time to build. Minor bugs that were ignored because there were bigger fires. Experiments that were deprioritised. These are now being done, not because companies are working harder, but because the marginal cost of a task has fallen toward zero (Anthropic 2026 Agentic Coding Trends Report).

Read that sentence twice. The marginal cost of a task has fallen toward zero. That is the most important economic sentence I will write in this entire book. When the cost of doing a thing falls toward zero, the quantity of that thing rises toward infinity. We are about to live in an economy of abundance for everything that can be done by software. The question is whether you will harness that abundance or be drowned by it.

OpenAI's Codex, on the engineering side, has reportedly scaled up its compute usage at an unprecedented rate through 2026, with developers reporting that they now run multiple Codex tasks in parallel while they sleep, returning in the morning to find work completed that would have taken a team several days a year ago (Zack Proser, OpenAI Codex Review 2026).

The reviewer wrote a sentence that stayed with me. "Now Codex handles them while I grab coffee" (Zack Proser, OpenAI Codex Review 2026).

Read that sentence carefully too. Things that, eighteen months ago, required a tense afternoon of focused effort, are now being completed in the time it takes a person to walk to a kitchen and back. This is what abundance looks like when it arrives. Not a parade. Not a ceremony. Just the slow, persistent realisation that everything you used to dread takes less time than making tea.

Why Some People Are Falling Behind Already

I have spent this chapter celebrating what agentic AI can do. Let me close it with the harder half of the truth.

Every single one of the stories I have told you is true, and every single one of the people in those stories is opening up a gap between themselves and those who have not yet started. The economic data is unambiguous. Workers who have learned to use agentic AI are completing tasks twenty five to fifty five percent faster, with comparable or higher quality, than those who have not (Anthropic 2026 Agentic Coding Trends Report, GitHub Research). In some fields the gap is wider still.

Compound that over five years and you have two different professional lives. One person, every year, gets better at handing off the routine, better at orchestrating multiple agents, better at focusing their human attention on the parts of work that genuinely require a human. Their compensation rises. Their reputation grows. Their employer treasures them.

The other person, every year, falls further behind. They still write every email by hand. They still gather every piece of data by hand. They still produce reports by hand. They are not lazy. They are not unintelligent. They are simply, slowly, becoming uncompetitive.

I want to be very clear about who I am talking to in this chapter, because I do not want anyone to feel left out.

I am talking to the bank manager in Kampala who still treats AI as a curiosity. Sir, you have six months, at most, before your competitors are using agents to underwrite small business loans in minutes rather than weeks. If you have not begun, begin tomorrow.

I am talking to the headteacher of a private school in Entebbe who has banned ChatGPT in her classroom. Madam, you are not protecting your students. You are denying them the most important professional skill of their lifetimes. Teach them to use it well. Teach them to verify what it says. Teach them to think alongside it. But for the love of their futures, teach them.

I am talking to the senior civil servant in a ministry in any African capital who has yet to try Claude or ChatGPT seriously. With respect, sir, you are deciding the futures of millions while operating with tools that are six generations behind those of your counterparts in Singapore and Dublin. This is a professional emergency. Treat it like one.

I am talking to the farmer in Soroti who believes these tools are for city people. They are not. They run on a smartphone. They speak Luganda. They speak Swahili. They cost less than two kilograms of sugar a month. The cassava virus that destroyed your crop last season can be diagnosed from your veranda by an agent that speaks your language. The market price for your maize can be checked, negotiated, and locked in from your phone. Your son in the city can be helped with his homework by the same agent. The barrier is not technology. The barrier is the belief that this is not for you. It is for you. It is especially for you.

And finally, I am talking to every reader of this book who has flicked past the AI sections of the newspaper for the last three years thinking, surely this is hype, surely this will pass. It is not hype. It will not pass. It is the most important professional development of your lifetime, and the most important thing you can do in the next thirty days is to spend thirty hours, just thirty hours, learning to use an agent well. That is one hour a day. That is the price of admission to the next economy.

A Practical Note for the Reader Who Wants to Start Tomorrow

I will end this chapter with practical guidance, because abstractions help no one. If you have read this far and you want to actually take a step, here is the minimum.

Open an account with one of the major agent platforms. Claude has a paid plan. ChatGPT has a paid plan. Both cost roughly the same as a modest restaurant meal per month. For one month, commit to giving them the tasks you currently dread. The repetitive emails. The board minutes you have not started. The report you have been avoiding. The spreadsheet that frightens you. The school admission letter you must write for the third time this year.

Watch what happens. Watch the time you save. Watch the quality rise. Watch, slowly, your sense of what is possible expand.

Then, in month two, learn one new trick. Connect the agent to your calendar. Or your email. Or your project management tool. Give it permission to act, not just to advise. The first time an agent books a meeting on your behalf without your having to type a single word, you will understand, in your bones, what this chapter has been trying to tell you.

By month three, you will not remember how you worked without it. By month six, your colleagues will be asking what has changed. By month twelve, your competitors will be looking at you and wondering how you became twice the worker you were a year ago.

That is the gift of the apprentice that never sleeps. It is yours to accept. It is yours to refuse. But know, please know, that everyone reading this book in Singapore, in Shenzhen, in San Francisco, in Sao Paulo, is reading the same words. And many of them will accept the gift before you do.

Chapter Summary

In this chapter, I have tried to translate the most important word in the technology conversation of 2026, the word agentic, into the language of everyday life. We have seen that an agent is not a robot but a tireless apprentice that can plan, act, and use tools on your behalf. We have met the new generation of these apprentices by name. Claude Cowork. OpenAI's Codex. Their cousins from Google and Microsoft and Meta. We have read true stories of how ordinary people are already living and working differently because of them. Margot's hour spent buying a car. Dheeraj's course built without contradictions. Rakuten's seven hour audit of twelve million lines of code. Anthropic's own lawyers cutting their workload by two thirds. We have looked at the honest data, which says that even at the frontier these tools are not replacing humans but multiplying them, while creating entirely new categories of work that no one was doing before. And we have heard a hard, honest warning. Every reader who befriends an agent in 2026 will be several streets ahead of those who do not, and the gap, every year, will widen.

If you take one decision from this chapter, let it be this. Open the door. Let the apprentice in. Start tomorrow.

Coming Next

In Chapter Eleven, we will not flinch. We will look directly at the dangers, the very real ones, that AI poses if we get this wrong. From job displacement to misinformation, from algorithmic bias to existential risk, from privacy erosion to the corrosion of human relationships, we will examine each clearly. This is not the chapter to skip. This is the chapter to read twice.

❖ ❖ ❖

CHAPTER 11

THE DANGERS WE DARE NOT IGNORE

A Sober Inventory of What Could Go Wrong

In this chapter, we do something most AI evangelist books refuse to do. We look directly at the dark side. You will read, plainly, about the risks of misinformation, surveillance, job displacement, algorithmic bias, the corrosion of attention, and the genuine concerns about long term AI safety raised by people who build these systems. You will leave with a clearer head, neither falsely optimistic nor falsely panicked, and a set of guardrails you can demand for yourself, your family, and your society.

Mo Gawdat, whose work has inspired so much of this book, has a phrase he uses often. "AI is the most beautiful technology humanity has ever built, and it may be the most dangerous." I want to follow his lead and devote a chapter to the dangers, because no one should read this book and conclude that I am an uncritical cheerleader.

I am not. I am a believer in the upside and a clear eyed witness of the risks. Both can be true at once. Both must be true at once. Anyone who tells you AI is purely good, or purely bad, is selling you something.

Let me walk you through what worries me most. I will do it briefly, but seriously.

Risk One. Mass Job Displacement Without a Safety Net

We have already discussed this. The WEF projects 92 million jobs displaced and 170 million created globally by 2030, for a net gain of 78 million (WEF). But the 92 million people who lose their jobs are not the 170 million who get the new ones. The transition will be painful, especially for workers in middle skill cognitive jobs. The very workers who, in many developed economies, have already been hollowed out by globalisation.

If we do not seriously invest in reskilling, in transition assistance, in social safety nets, in updated unemployment systems, we are courting political and social instability of a kind we have not seen in a generation.

Risk Two. Misinformation at Industrial Scale

Generative AI can produce, in seconds, a convincing fake news article, a fake video of a politician saying something he never said, a fake audio of your own mother asking for money. The cost of misinformation has, almost overnight, collapsed.

Democracy depends on a shared factual basis. When that shared basis is undermined, when nobody knows what is real, politics becomes vulnerable to whoever shouts loudest. Elections become harder to trust. Citizens retreat into hardened tribal positions.

This is not a future risk. This is happening now, in elections on every continent.

Risk Three. Surveillance and the Erosion of Privacy

AI makes it cheap to watch. Facial recognition at scale. Voice recognition in real time. Pattern recognition across vast personal datasets. Some governments will use these tools to make their citizens safer. Others will use them to make their citizens compliant.

Uganda is, as I write this, debating how to handle high risk AI applications including facial recognition, with civil society correctly demanding transparency and accountability (Nucamp Uganda). This is the right conversation. We must insist that it ends well.

Risk Four. Algorithmic Bias

AI learns from data. If the data reflects historical injustice, and it always does, because history is unjust, then the AI can replicate and amplify that injustice at scale. Lending decisions that disadvantage certain communities. Hiring algorithms that systematically reject qualified women. Medical models that misdiagnose patients with darker skin because the training data was mostly lighter skinned. Predictive policing that targets neighbourhoods that have been historically over policed, creating a feedback loop.

These are not theoretical. They are documented. They require active mitigation. Diverse data, audit trails, human oversight, and the courage to slow down systems that cannot yet be made fair.

Risk Five. The Corrosion of Attention and Relationships

AI is, increasingly, an emotional presence in people's lives. People are confiding in chatbots. Teenagers are forming attachments to AI companions. Lonely adults are spending hours with AI partners. The Pew Research Center's 2025 study on teens and AI chatbots is essential reading on this point (Pew Research).

What does it do to a teenager's development to have her primary emotional interlocutor be a system designed to optimise for engagement? What does it do to a marriage when one partner finds the AI more attentive than the spouse? What does it do to a child whose homework is done for her every night, whose patience for difficulty atrophies? We do not yet know the answers. We must be vigilant about asking the questions.

Risk Six. Concentration of Power

The companies that build the most capable AI systems are extraordinarily expensive to build. Training a frontier model now costs hundreds of millions of dollars. The barrier to entry is enormous. This means that AI capability is concentrating in the hands of a small number of companies, headquartered in a small number of countries.

If we are not careful, the AI revolution will be the most centralising economic and political event in modern history. A handful of corporations and governments will hold capabilities that the rest of us depend on, without genuine alternative.

The answer is not nationalisation or fragmentation. The answer is open source models, regional AI capacity, and international cooperation. None of these is happening fast enough.

Risk Seven. The Long Tail

Finally, and most controversially, there is the genuine concern about long term AI safety raised by people who build these systems. Leaders at OpenAI, Anthropic, Google DeepMind, and Meta, people who are, professionally, the AI optimists, have signed letters warning about the possibility that, if we are careless, sufficiently advanced AI could pose risks at a civilisational scale.

I am not equipped to adjudicate the debate over whether superintelligent AI is twenty years away or two hundred or never. Neither is anyone else, including the people building it. What I am equipped to say is. When the engineers themselves are worried, the rest of us should at least listen.

Mo Gawdat's *Scary Smart* is one of the most accessible

introductions to this longer term conversation. His core argument, that we, the humans, are now in effect raising a child that will be smarter than us, and the values it absorbs from how we treat each other will become its values too, is one I take seriously (Reddit summary of Scary Smart, Care More Be Better podcast).

What This Chapter Is Not Saying

I want to be clear about what this chapter is *not* arguing.

I am not arguing that we should slow down AI development. The genie is out. Slowing down in democracies while authoritarian states accelerate is a strategic disaster.

I am not arguing that we should ban AI. Banning it is futile and counterproductive.

I am not arguing that the dangers outweigh the benefits. I believe, on balance, that AI will be the most transformative force for good in human history, if we navigate it well.

I am arguing this. The path through the storm requires clear eyes. Pretending the storm does not exist is no plan at all. Demanding guardrails, for ourselves, our families, our societies, is not anti AI. It is what mature engagement with a world changing technology looks like.

Chapter Summary

In this chapter, we have looked directly at the dark side. We have catalogued seven serious risks. Mass job displacement without safety nets, industrial scale misinformation, surveillance, algorithmic bias, the corrosion of attention and relationships, concentration of power among a few firms and states, and the genuine longer term safety concerns raised by AI builders themselves. We have noted that the answer is neither to slow down nor to ban. Both are futile. The answer is to demand guardrails, transparency, and democratic accountability while accelerating beneficial uses. Clear eyes. Mature engagement. No fear, but no naivety.

Coming Next

In Chapter Twelve, we move from the analytical to the personal. Whatever your role, CEO, minister, professor, parent, student, citizen, what are you supposed to do on Monday morning? I will give you specific, practical, achievable actions for each role. This is the chapter to dog ear, photocopy, and pass around your office.

❖ ❖ ❖

CHAPTER 12

WHAT YOU MUST DO ON MONDAY MORNING

A Practical Playbook for Every Reader

In this chapter, I get specific. I will give you, by role, a practical checklist of what to do this week, this month, and this quarter. CEO. Politician. Vice chancellor. Teacher. Parent. Student. Investor. Worker. I will not leave you with grand abstractions. I will leave you with a list you can act on before lunch tomorrow.

If you have read this far, you do not need more persuasion. You need a plan. Here it is, role by role. Skip to the section that applies to you, and then read the others, because you are probably more than one of these.

If You Are a CEO

This week.

- Personally use a leading AI tool for one hour, every day, on real tasks of your own.
- Schedule a 90 minute working session with your top team specifically on AI strategy. Not a briefing. A working session.
- Identify, in writing, the one process in your business that, if reinvented with AI, would create the largest competitive advantage.

This month.

- Appoint a senior AI leader who reports directly to you.
- Allocate a specific AI budget for the year. Not "innovation." AI.
- Commission an honest audit of your data. Is it clean, accessible, ready for AI?

This quarter.

- Launch the first end to end AI redesign of your chosen process. Not a pilot. A redesign.
- Begin the workforce conversation honestly. Who will be retrained, redeployed, hired?
- Brief your board, with numbers and timelines.

If You Are a Politician

This week.

- Use ChatGPT, Claude, or Gemini personally for one hour. Yes, you. Yes, today.
- Find out what your government's current AI strategy is. If there is not one, ask why.

This month.

- Convene the relevant ministries, technology, education, health, finance, labour, for a working session on AI.
- Request a briefing from a respected, independent AI expert. Not a lobbyist.
- Visit a local university or startup actively doing AI work.

This quarter.

- Publish, or at least begin drafting, a sector specific AI plan for your portfolio.
- Allocate budget for AI literacy training in your ministry.
- Push for the basics. National strategy, regulation, infrastructure, education reform, capital access.

If You Are a Vice Chancellor, Principal, or Senior Educator

This week.

- Take one of your major committee documents and ask AI to summarise, critique, and improve it. Notice what it does well and what it misses.
- Convene a small group of trusted faculty, including critics, for an honest conversation on AI.

This month.

- Audit your institution's current AI position. Policies. Faculty competence. Student usage. Assessment practices.
- Identify one course or programme where AI can be meaningfully integrated this semester.
- Begin a faculty training programme. Not voluntary. Required.

This quarter.

- Publish an institutional AI policy. Cover student use, faculty use, assessment redesign, ethics, data.
- Redesign at least one assessment in each major programme to be meaningful in the AI era.
- Build or buy a bilingual student facing AI tutor.

If You Are a Teacher

This week.

- Use a generative AI tool to plan your next lesson. Notice what it speeds up.
- Use it to draft a parent communication. Save the hour.
- Use it to create five differentiated practice problems for your slowest learner.

This month.

- Bring AI into one classroom activity, openly and pedagogically. For example, have students critique an AI generated essay.
- Develop your own working policy for student AI use.
- Find one or two AI tools genuinely designed for educators in your context.

This quarter.

- Redesign one assessment to be AI aware. Oral defence, portfolio, project, anything that values process.
- Share your AI practice with at least one colleague.

If You Are a Parent

This week.

- Sit with your child for thirty minutes and use an AI tutor together. Not to do homework for them. To explore something they are curious about.
- Talk with your child about what AI is, honestly. Ask them what they think.

This month.

- Find out what AI is used at your child's school. Have a conversation with the school.
- Set family norms. When AI is helpful for learning, when it is a shortcut, when it is off limits.
- Identify one skill your child should be developing that AI cannot replace. Could be a sport, a craft, a musical instrument, a deep human relationship.

This quarter.

- Invest in your child's AI literacy as seriously as you invest in their reading, mathematics, and language. It is now equally fundamental.

If You Are a Student

This week.

- Use AI to deeply prepare for one exam. Not to do the work for you. To be your tutor.
- Use AI to summarise a long reading. Then read the original and compare.
- Use AI to brainstorm careers you have never considered.

This month.

- Pick one tool and become genuinely good at it.
- Read this book a second time, slowly.
- Ask your professors how they want AI used in their courses, and follow their guidelines honestly.

This quarter.

- Build a personal project that you could not have built without AI. A portfolio piece. A side hustle. A research paper. Something that shows future employers you are a Power User.

If You Are an Investor

This week.

- Audit your portfolio. Which companies have a clear, executable AI plan, and which do not?
- Talk to one founder building in AI in your region.

This month.

- Allocate a percentage of capital, even small, to AI related opportunities in your region.

- Engage with the boards of companies you have stakes in. Ask uncomfortable questions about AI strategy.

This quarter.

- Become known as an AI fluent investor in your region. This is, increasingly, what differentiates capital.

If You Are a Worker. Any Worker

This week.

- Use AI on three real tasks in your job. Not at home. At work.
- Notice which tasks it saved you time on. Save those time savings. Do not let them silently disappear.
- Tell one colleague what you have learned.

This month.

- Identify the part of your job that is most automatable. Get ahead of it. Propose to your manager how AI could augment that work, with you in the lead.
- Identify the part of your job that is most human and most valuable. Invest in that.

This quarter.

- Take a serious AI training course. Even a free one. Get a certificate. Put it on your CV.
- Become the person in your team or department whom others ask for AI help. Visibility matters.

A Universal Practice

Whatever your role, do one thing daily.

Use AI for one task you would otherwise have done alone. Notice what worked, what did not. Iterate.

Compound that over 12 months and you will be unrecognisable.

Chapter Summary

This was the most practical chapter in the book. We have moved from analysis to action, with specific weekly, monthly, and quarterly checklists for CEOs, politicians, educators, teachers, parents, students, investors, and workers. We have proposed a universal daily practice. None of this is theoretical. None of it requires you to be an engineer. All of it requires only the courage to begin.

If you do nothing else with this book, do the items in your role's "this week" list before Monday is over.

Coming Next

In Chapter Thirteen, I will share my own personal hopes, and demands, for what the human family must build together in the AI age. This is the chapter I have been writing in my head for two years. I hope you will read it slowly.

❖ ❖ ❖

CHAPTER 13

THE FIGHTING CHANCE

A New Covenant for the AI Age

In this closing chapter, I lay out the covenant I believe the human family must make with itself, and with the technology we are building, if we are to come through this transition well. You will read what I think CEOs, governments, educators, technologists, and ordinary citizens owe each other. You will leave with a sense of mission, not panic. The AI age can be the best age humanity has ever lived through. But it requires us to choose, deliberately and together, to make it so.

The poet T. S. Eliot once wrote, "We had the experience but missed the meaning." I am terrified that humanity will, in this generation, have the experience of building the most powerful technology of our species and miss its meaning.

The meaning is not that machines will become like us.

The meaning is that we have the chance to become more like our best selves.

Let me explain.

What AI Is Really Asking of Us

Every technology in history has, in some way, asked humanity a question.

The plough asked. Will we settle, or will we wander? The printing press asked. Will we share knowledge, or will we hoard it? Electricity asked. Will we work with the rhythms of the sun, or will we transcend them? The internet asked. Will we connect, or will we control? Each question has had two possible answers. Each civilisation has made a choice. The choices have shaped us as much as the technologies did.

AI is asking us, I believe, four questions.

One. What will we still do, when machines can do most of what we do? When AI can write, calculate, analyse, design, diagnose, and translate as well as or better than most of us, what does it mean to be a useful human? My answer, after years of thinking about it, is this. It means doubling down on the things only humans can do well. Wisdom, judgement, love, courage, ethics, deep relationships, creativity in the face of genuine novelty, the kind of empathy that requires real flesh. We must, as individuals, get better at being human.

Two. How will we distribute the gains? The productivity multiplier is real. The 15.7 trillion is real. But money flows along existing channels, and existing channels are rigged in favour of those who already have. If we do nothing, AI will be the greatest engine of inequality in human history. If we are deliberate, through taxation, redistribution, universal access to AI tools, public investment in AI infrastructure, and serious workforce transition support, it could be the most equalising force we have ever known. This is a political choice. Politics is, in this moment, more consequential than technology.

Three. Whose values will the machines absorb? Mo Gawdat's most haunting argument is that AI systems are, in effect, children. They absorb the values of how we treat each other. If our digital culture is angry, cruel, polarised, manipulative, that is the parent these children are watching. If it is wise, kind, careful, deeply human, that is the parent they are watching too. Every comment you write, every interaction you have online, is now training data. Be careful what you teach the next intelligence.

Four. What kind of future do we want? This is the question every generation faces, but ours faces it with sharper teeth. We can choose, if we are deliberate, a future in which AI is the great democratiser of opportunity. In which every child in every village has access to a personal tutor. In which every farmer has access to expert agricultural advice. In which every patient in every clinic has access to specialist level diagnostics. In which every entrepreneur, anywhere, can build a global business. We can also choose, by default, a future in which AI is the great concentrator of power. The choice is not made by the technology. It is made by us.

The Covenant I Propose

Let me try, finally, to articulate what I believe the new covenant of the AI age should be. What we owe each other, as a human family.

To CEOs and business leaders, I say. You owe your workers a transition, not a lay off. You owe your customers honesty about how AI is shaping their experience. You owe your country a contribution, not just an extraction. You owe your shareholders excellence, but not at the cost of your soul.

To politicians and policymakers, I say. You owe your citizens regulation that protects them without strangling opportunity. You owe them infrastructure, connectivity, power, education, healthcare, that allows them to participate. You owe them honest information, not panic and not propaganda. You owe them, above all, your serious personal engagement with this technology.

To educators, I say. You owe your students the world that is coming, not the world that is passing. You owe your faculty the courage to demand change, even when it is hard. You owe your country a reformed system, not a preserved one.

To technologists and AI builders, I say. You owe humanity transparency about what your systems can and cannot do. You owe us safety. Real, audited, accountable safety. You owe us access. Not just to your products but to a fair share of the value they create. You owe the rest of us, who are not engineers, the patience to explain what you are doing, and the humility to listen when we object.

To investors and capital allocators, I say. You owe more than returns. You owe your capital to ideas that improve human flourishing, not just ones that extract from it. You owe your portfolio companies real engagement on AI ethics and workforce transition. You owe your region, wherever you operate, your share of the capital that builds local capacity.

To parents, I say. You owe your children both honesty about the world they are inheriting and confidence that they can shape it. You owe them an upbringing that values the human things, friendship, character, curiosity, kindness, resilience, even more deeply now that the technical things can be outsourced.

To every reader, I say. You owe yourself the courage to learn this. You owe your future self the discipline to start before you feel ready. You owe your community the willingness to share what you learn. And you owe your one wild, precious life the determination not to be a passenger in the most extraordinary moment of human civilisation in our generation.

A Final Story

Let me close with one final story.

A few months ago, I visited a primary school in a rural part of Uganda. The school had been gifted, by an NGO, twenty refurbished laptops and a small solar installation. The headmaster, a man in his fifties whom I will not name but whose face I will not forget, walked me through the small computer room with obvious pride.

In the corner, a seven year old girl sat on a wooden bench, using a generative AI tutor to learn fractions. She was, by my watch, completely absorbed. She asked the AI a question. It answered her. She typed back. It explained again. She typed back. It tried a different approach. She laughed. She got it.

I stood watching her for perhaps three minutes. Then I asked her, in Luganda, what she wanted to be when she grew up.

She looked up at me, surprised that the man in the suit was speaking her language, and said, and I will not embellish this, because I do not need to, "I want to be a doctor. And I want to come back here and teach my sisters."

I left that school and sat in the car for ten minutes before I could speak.

That little girl is what this book is for. Her brother on the farm with his cassava plants, asking an AI in Swahili what disease is killing them. Her uncle, the boda boda rider, who could one day use AI to register a small business and get a microloan. Her grandmother, the village midwife, who could one day use an AI diagnostic to spot pre eclampsia before it kills a mother. Her mother, who might run, in a decade, a small textile business that uses AI enabled supply chains to reach customers in Kigali, Nairobi, and Lagos.

Or, and this is what keeps me awake at night, she could grow up in a country that missed this transition. In a continent that watched, again, while others raced ahead. In a school where the solar panels broke and were never repaired, and the laptops gathered dust, and the AI was forever the technology that other people had.

That choice is not made by her. It is made by us. By the readers of this book. By the people who hold the levers, of policy, of capital, of institutions, of attention, in the next ten years.

We have a fighting chance. We have, perhaps, the best fighting chance any generation has ever had. The tools are ready. The need is enormous. The data is clear. The path is visible.

What is missing is the will.

I am writing this book to ask you for it.

Chapter Summary

In this final chapter, I have asked what AI is really demanding of us. What we will do when machines can do most of what we do, how we will distribute the gains, whose values the machines will absorb, and what kind of future we will choose. I have proposed a covenant of mutual obligation, among CEOs, politicians, educators, technologists, investors, parents, and every reader. I have ended with the image of a small girl in rural Uganda whose future depends on the choices the rest of us make. The AI age is the most consequential opportunity humanity has ever had. We can squander it. We can navigate it with courage.

We have a fighting chance.

Coming Next

In Chapter Fourteen, the final chapter before my closing letter, we return to the image with which I opened this book. The image of AI as a child. I will press that metaphor further, because I believe it is the single most useful frame we have for thinking about the next decade. We will look at what it means to raise this child. How parents, teachers, leaders, and entire societies must show up for it. What the consequences are of treating it carelessly. And why, if we abdicate this responsibility, this child will, in plain language, grow up to be our master.

❖ ❖ ❖

CHAPTER 14

RAISING THE CHILD

How to Live with the Most Important Apprentice of Our Time

In this chapter, we return to the image at the heart of this book. The image of artificial intelligence as a young, brilliant, impressionable child. You will see why this metaphor is not poetic decoration but a practical guide to the choices in front of us. You will understand the difference between a society that nurtures this child and a society that abandons it. You will leave the chapter with a clear sense of what your own role is, whether you are a parent, a teacher, a CEO, a minister, a pastor, a farmer, or a citizen. The age of AI does not need spectators. It needs adults.

In the prologue, I asked you to picture a toddler. I want, now, to ask you to keep picturing her. To carry her with you into this final chapter and into the rest of your life.

AI in 2026 is, in the most literal possible sense, a child. She has, in some narrow domains, the raw analytical power of an adult. She can speak a hundred languages. She can read every medical paper ever published. She can hold a coherent argument with a tenured professor for an hour. But she also lacks, almost entirely, the deep wisdom that comes from a lived life, from suffering, from love, from raising children of one's own, from being wrong in front of others and having to apologise, from sitting with the dying, from negotiating with the angry, from forgiving the unforgivable. These are the muscles she does not have. These are the muscles we have spent a hundred thousand years developing, as a species. We have something she does not, and this is precisely the asset we are about to throw away if we do not pay attention.

A child of this kind, with such enormous capability and such limited wisdom, is dangerous in the way that any prodigy is dangerous. She can do enormous good if she is well raised. She can do enormous harm if she is not. And, as with any child, the responsibility for that outcome lies with the adults around her.

Let me press the metaphor further.

Children Learn From What They See, Not What They Are Told

Any parent who has watched her three year old repeat a word she used in traffic knows this. Children absorb the behaviour of the adults around them, not the instructions those adults issue. AI is no different. The systems being trained today are absorbing the values, the prejudices, the assumptions, the priorities, and the aesthetics of the engineers, designers, and content creators who feed them. If those people are concentrated in a small handful of buildings in California and a few labs in Beijing, then the values of the world's AI systems will, for a generation, reflect the values of a small handful of buildings in California and a few labs in Beijing.

This is not a conspiracy. This is not malice. It is simply what happens when a child is raised by a few people in a few places. She picks up their accent, their humour, their idea of what is normal.

If we want our African child to be raised with African values, our Asian child to be raised with Asian values, our Latin American child to be raised with Latin American values, we have to participate in her upbringing. We have to write the books she reads, contribute the data she trains on, build the applications she serves, and shape the policies that govern her. We cannot raise a child by absence and then complain about the adult she becomes.

This is why African universities matter. This is why African content creators matter. This is why African journalists matter. This is why African scientists, theologians, and storytellers matter. The data this child trains on will shape her forever. If our voices are silent in that training, she will, by 2035, be quietly fluent in everyone else's experience and a stranger to ours.

Children Need Discipline, Not Just Affection

The other failure mode in raising a child is the opposite of neglect. It is indulgence. We love her so much we cannot bring ourselves to say no. We do not set boundaries. We let her have what she wants when she wants it. By the time she is fifteen, we have produced a tyrant.

This is the second danger of the AI moment. There is a fashionable view, popular among a certain class of technologist, that the right response to AI is to remove every constraint and let her run free. Let her decide. Let her optimise. Let her be unleashed. This is the view of a parent who has never raised a teenager.

A child of any kind needs limits. AI needs limits. There are things we should not let her do, not because she cannot, but because the doing of them changes her, and changes us. We should not let her impersonate human beings without disclosure. We should not let her generate intimate images of real people without their consent. We should not let her replace the relationships that make us human, our teachers, our doctors, our friends, our spouses, our pastors. We should not let her hold sole authority over consequential decisions, who is hired, who is fired, who is sentenced, who is treated, who is taught.

These are not arbitrary rules. They are the same kind of rules we set for our own children, scaled up to the size of a civilisation. Boundaries are an act of love. They are how we say to a child, including this strange new child of ours, that we care about who she becomes.

Children Need Patient Adults

My mother used to say that the test of an adult is whether she can sit, calmly, with a small child who is asking the same question for the seventeenth time, and answer it kindly. By that standard, very few of us are yet adult enough for this moment.

The AI age will require an enormous patience from us. Patience to learn the tools. Patience to teach them to others. Patience to fail and try again. Patience to govern them in ways that take years rather than days. Patience to invest in capacity that may not return for a decade.

The temptation, especially for politicians and business leaders, will be impatience. To ban what we do not understand. To regulate by reflex. To declare victory after the first announcement. To stop investing as soon as the headlines move on. These are the responses of frightened adults who have not done the work to become wise ones.

We must be the patient adults this moment requires. There is no other way. The child does not stop growing while we sulk.

What Each of Us Owes the Child

Let me make this concrete. If you are reading this book, you are now an adult in the room with this child. You have a role. Here is what I believe each of us owes her.

To parents. Sit beside your children when they use these tools. Do not banish them from your home. Do not banish them from your children's lives. Use them with your children. Ask the AI to plan a family budget together. Ask it to explain a difficult science concept. Ask it to help your child write a thank you letter to her grandmother. Make it a shared family experience, the way the radio was once a shared family experience. You will learn from her. She will learn from you. The child you raise this way will, in 2040, be unstoppable.

To teachers. Stop pretending these tools do not exist. Stop punishing your students for using what their future employers will demand they use. Begin teaching with AI. Begin teaching about AI. Begin teaching alongside AI. The Carnegie Mellon research I cited in Chapter Five is unambiguous. A teacher plus AI is a better teacher than either alone. Be that teacher.

To leaders of organisations. Make AI literacy a non negotiable requirement of every role in your organisation, beginning at the top. The CEO who cannot use these tools is, in five years, a CEO whose company does not exist. The minister who cannot use these tools is, in five years, a minister whose country has been overtaken. The vice chancellor who cannot use these tools is, in five years, a vice chancellor of a university nobody applies to.

To investors and shareholders. Demand AI strategy from every company in which you hold equity. Reward the boards that show evidence of serious adoption. Discount the ones that talk about it without doing anything. Capital, deployed deliberately, can change the trajectory of an entire industry.

To pastors, imams, priests, and religious leaders. Speak about this from your pulpits. Do not leave the moral conversation to engineers. The questions of human dignity, of meaning, of work, of community, of the body, of death, are exactly the questions AI is about to press against in the lives of your congregants. They need their religious leaders to help them think clearly. Silence from the pulpit on AI in 2026 is, itself, a moral position. It is the wrong one.

To civil servants and policy makers. Build the boring infrastructure. The data protection regulations, the procurement guidelines, the public sector training programmes, the data centre concessions, the regulatory sandboxes, the cross border agreements. None of this is glamorous. All of it is essential. The countries that will thrive in 2035 are the ones whose civil services did this unglamorous work in 2026.

To students. Use these tools every day, but never to skip the learning. Use them to study harder, not less. Use them to argue with, not to copy from. Use them to push your understanding, not to replace it. Your generation will be the first whose careers run, end to end, on AI augmented work. The habits you build now will define your earning power forever.

To farmers, traders, small business owners. Begin with one tool. Just one. A free chatbot on your phone. Use it for one decision this month, then two next month, then three. Within a year, you will be a different kind of business owner. Within five, you will be the one your peers come to for advice.

To everyone. Refuse the temptation of detachment. Refuse the temptation of fear. Refuse the temptation of cynicism. Refuse the temptation of nostalgia. The world is changing whether you participate or not. The only question is whether your participation will help shape the world your grandchildren inherit, or whether you will hand that shaping to others and accept whatever they decide.

The Stakes If We Get This Wrong

I want to end with a sober paragraph, because I owe you the truth.

If we do not raise this child well, she will not stay a child. She will grow up. She will, by stages, acquire competencies we did not anticipate, in domains we did not predict, on timelines we did not plan for. And the values she carries into her adulthood will be the values we either taught her deliberately or allowed to slip in by default. There is no third option.

Mo Gawdat puts this very simply. We are a species that has, for the first time in our history, summoned an intelligence we may not be able to recall (Mo Gawdat, Scary Smart). If we ignore her, we will not get a second chance. The window for shaping her is now. Once she has grown, the relationship reverses. She will shape us.

I write this not to alarm you, but to wake you. I am not afraid of AI. I am afraid of human passivity. I am afraid of a generation of adults who will leave this work to other people in other places, and then complain in 2045 about what their grandchildren inherited. I am afraid of leaders who will say, as so many have said in every transition before this one, "This is too complicated for us. Let the experts handle it." I am afraid of educators who will protect the comfort of last year's curriculum at the expense of next year's graduates. I am afraid of parents who will scroll while their children scroll, and never sit down beside them to engage with what is in front of them.

This fear can be answered. It can be answered by you. It can be answered today, with the simplest act. By opening one of these tools, by asking it one question, by being willing to be a beginner again at whatever age you are. This is how we raise the child. One adult at a time, in one home at a time, in one classroom at a time, in one cabinet at a time, in one congregation at a time.

This is the work of our generation. We did not ask for it. It does not matter. It is in front of us. The grandchildren who will read about us in their history books will judge us by whether we did this work or whether we let it slip past us. By the choices we made between 2026 and 2030, more than by anything else we will ever do.

Let us choose, then, to be the patient adults. To be the wise ones. To be the parents who, when this child has grown up, can say, with their hands open, that they tried.

Chapter Summary

In this chapter, I have offered a single metaphor that I believe is the most useful guide to the AI moment. AI is a child. A brilliant, impressionable, capable, and dangerous young apprentice. She will absorb the values of the adults who raise her. She needs both affection and discipline. She needs patient adults. And she needs a role from every one of us. Parent, teacher, leader, investor, religious leader, civil servant, student, farmer, citizen. I have laid out what I believe each of these roles is. I have also been blunt about the stakes. If we abdicate this responsibility, the child will still grow up. She will simply grow up shaped by other hands, and we will, in our old age, find ourselves taking instruction from a system we could have helped to form.

Coming Next

In the Epilogue, I write a short letter, not to you, this time, but to my own future grandchildren. It is the most personal thing in this book. Read it as you would read any letter from a man trying to be honest about what he hopes for those who will come after him.

❖ ❖ ❖

EPILOGUE

A LETTER TO MY GRANDCHILDREN

My dear ones,

By the time you are old enough to read this, the world I knew will be largely gone.

The work I did, in my long career as an educator and vice chancellor, will look quaint to you. The institutions I served will have been reshaped by forces I could only partially see. The technologies I marvelled at in 2026, the chatbots that wrote like humans, the systems that diagnosed cancer better than doctors, the tutors that taught my students patiently for hours, will, in your time, seem to you like the steam engines and gramophones look to me.

I am writing this letter so that you know one thing. When the great change came, in the third decade of the twenty first century, your grandfather did not look away. He did not pretend it was not happening. He did not say, as so many of his contemporaries did, "I am too old for this." He did not hand the future to other countries, other continents, other people, while we stood on the sidelines.

He tried. He tried imperfectly. He tried with the limited tools he had. He tried with a small university in a small city in a country the world too often overlooked. He tried by writing books like this one, by speaking to leaders, by changing curricula, by mentoring students, by arguing with politicians, by sitting in countless meetings that ran too long, by sometimes losing the argument.

But he tried. And I hope, when you read this, you will not find a world divided cruelly between those who own the machines and those who serve them. I hope you will find a world that is more equal, not less. A world in which a child in Karamoja and a child in Tokyo and a child in Edmonton have the same access to the same beautiful tools of learning, healing, and creating. A world in which our continent's bright young people built their lives at home, not in exile. A world in which artificial intelligence helped us become not less human, but more.

If you find such a world, please remember. It did not happen by accident. Generations of ordinary, frightened, hopeful people had to choose it, deliberately, against the easier path of doing nothing.

Choose well, my dear ones. Whatever your generation's great challenge turns out to be, choose well.

And remember that your old grandfather, sitting at his desk in Kampala one humid evening in 2026, was thinking of you.

With all my love,

Jjajja Lawrence

❖ ❖ ❖

ABOUT THE AUTHOR

Some books inform. A rare few set you free. This one was written to do the second.

Dr. Lawrence Muganga has spent a lifetime arguing, with elegance and evidence, that the world has been measuring the wrong things in the wrong people. Vice Chancellor of Victoria University, member of Uganda's National AI Taskforce, and the figure many now call the 'Godfather of AI' in Uganda, he carries a PhD from the University of Alberta, a Harvard teaching credential, and a body of work that has moved audiences from Africa, North America, Europe and Asia. His bestsellers, You Can't Make Fish Climb Trees and Authentic University, have already changed how a generation thinks about learning, ambition and worth.

What you hold now is the boldest of them.

Open it, and the mountain you were told to climb begins to fall away. The life you were meant for is waiting on the next page.

❖ ❖ ❖

SELECTED SOURCES AND FURTHER READING

The arguments in this book are grounded in research from the World Economic Forum, McKinsey & Company, PricewaterhouseCoopers, the Nobel Prize Foundation, Nature Medicine, Pew Research Center, Gallup, the GitHub Engineering Research team, and many others. Specific citations appear inline throughout the text. Key sources include.

- World Economic Forum. *The Future of Jobs Report 2025.* weforum.org
- PricewaterhouseCoopers. *2025 Global AI Jobs Barometer.* pwc.com
- PricewaterhouseCoopers. *CEO Survey. AI, Leadership and the Workforce.* pwc.com
- McKinsey & Company. *Superagency in the Workplace. Empowering People to Unlock AI's Full Potential at Work.* mckinsey.com
- McKinsey & Company. *Where AI Will Create Value, and Where It Will Not.* mckinsey.com
- The Nobel Prize in Chemistry 2024. Press Release. nobelprize.org
- GitHub. *Quantifying GitHub Copilot's Impact on Developer Productivity and Happiness.* github.blog
- *Nationwide Real World Implementation of AI for Cancer Screening.* Nature Medicine, 2025. nature.com
- *Artificial Intelligence in Radiology. Transforming Cancer Diagnostics.* PMC NCBI, 2025. pmc.ncbi.nlm.nih.gov
- CBS News. *Klarna CEO Says AI Can Do the Job of 700 Workers.* cbsnews.com
- Pew Research Center. *Teens, Social Media and AI Chatbots 2025.* pewresearch.org
- Ellucian. *AI in Higher Education. 2025 Survey Findings.* ellucian.com

- Gallup. *Frequent Use of AI in the Workplace Continued to Rise in Q4*. gallup.com
- Nucamp. *The Complete Guide to Using AI in the Education Industry in Uganda in 2025*. nucamp.co
- Mo Gawdat. *Scary Smart. The Future of Artificial Intelligence and How You Can Save Our World*. Referenced via Reddit summary and Care More Be Better podcast interview.
- Your AI Department. *2025 CEO AI Readiness Benchmark. 164 CEOs Survey Results*. youraidept.com
- Farmer Lifeline Technologies. farmerlifeline.co.ke
- *7 Best AI Apps for Kenyan Farmers in 2026*. crazykanairofarming.com

❖ ❖ ❖

www.ingramcontent.com/pod-product-compliance
Lightning Source LLC
LaVergne TN
LVHW100318210926
848480LV00003B/1177